CRTD-Vol. 79

I0031380

Technical Peer Review Report
Report of the Review Panel

Yucca Mountain: Waste Package Closure Control System

ASME International

Institute for Regulatory Science

DISCLAIMER

This report was prepared through the collaborative efforts of the American Society of Mechanical Engineers (ASME) Center for Research and Technology Development and the Institute for Regulatory Science (referred to thereafter with the collaborators as the Society) for the Office of Science and Technology Development of the U.S. Department of Energy (referred to hereafter as the Sponsor).

Neither the Society nor the Sponsor, or others involved in the preparation or review of this report nor any of their respective employees, members, or persons acting on their behalf, make any warranty, expressed or implied, or assume any legal liability or responsibility for the accuracy, completeness, or usefulness of any information, apparatus, product, or process disclosed, or represent that its uses would not infringe privately owned rights.

Information contained in this work has been obtained by the American Society of Mechanical Engineers from sources believed to be reliable. However, neither ASME nor its authors or editors guarantee the accuracy or completeness of any information published in this work. Neither ASME nor its authors and editors shall be responsible for any errors, omissions, or damages arising out of the use of this information. The work is published with the understanding that ASME and its authors and editors are supplying information but are not attempting to render engineering or other professional services. If such engineering or professional services are required, the assistance of an appropriate professional should be sought.

Statement from By-Laws: The Society shall not be responsible for statements or opinions advanced in papers . . . or printed in its publications. (7.1.3)

For authorization to photocopy material for internal or personal use under circumstances not falling within the fair use provisions of the Copyright Act, contact the Copyright Clearance Center (CCC), 222 Rosewood Drive, Danvers, MA 01923, Tel: 978-750-8400, www.copyright.com. Requests for special permission or bulk reproduction should be addressed to the ASME Technical Publishing Department.

TABLE OF CONTENTS

Preface

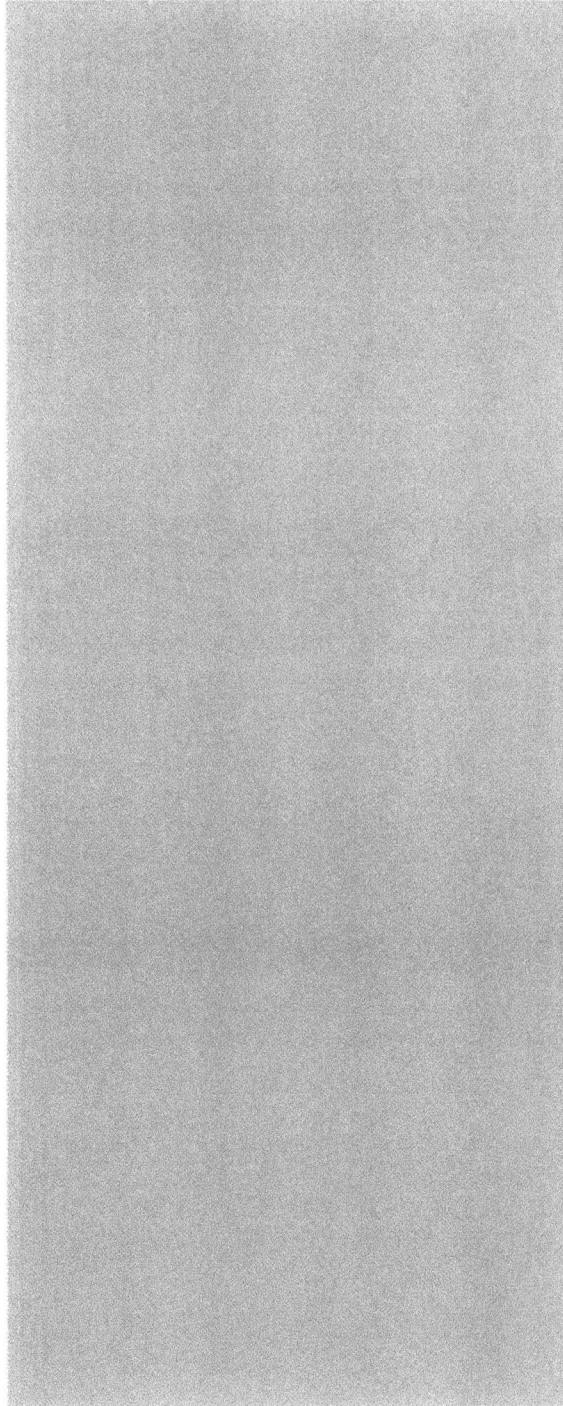

This report contains the results of a peer review performed jointly by the American Society of Mechanical Engineers (ASME) and the Institute for Regulatory Science (RSI). Based on a request from the Idaho National Engineering and Environmental Laboratory of the U.S. Department of Energy (DOE), a Review Panel (RP) was established to peer review "Yucca Mountain - Waste Package Closure Control System". Consistent with the ASME/RSI process: 1) a Type III review was requested; and 2) the following RP was appointed by the Peer Review Committee for Energy and the Environment (PRCEE) of ASME:

Richard Adams
George Cook, Ph.D., Chair
Francis Patti
Karyanil Thomas

During the period covered by this report, the ASME PRCEE overseeing the peer review consisted of the following individuals:

Charles O. Velzy, Member of Executive Panel (EP) and Chair
Ernest L. Daman, Member of EP
Nathan H. Hurt, Member of EP
A. Alan Moghissi, Member of EP; Principal Investigator of the Peer Review Program
Gary A. Benda
Erich W. Bretthauer
Irwin Feller
Robert A. Fjeld
William T. Gregory, III
Peter B. Lederman
Jeffrey A. Marqusee
Lawrence C. Mohr, Jr.
Goetz K. Oertel
Glen W. Suter, II
Cheryl A. Trottier

The supporting staff included the following individuals:

Michael Tinkelman: Director of Research at the Center for Research and Technology Development of ASME in Washington, DC; and Administrative Manager of the ASME PRCEE.

Betty R. Love: Executive Vice President, RSI, Columbia, MD; and Administrative Manager of the Peer Review Program.

Sorin R. Straja: Vice President for Science and Technology, RSI; and Technical Secretary.

M.C. Kirkland: Vice President for Southeastern Region, RSI.

Sharon D. Jones: Director of Training Programs, RSI; and Manager of Review Panel Operations.

The biographical summaries of the members of the RP, the PRCEE, and the technical staff are located at the end of this report.

The Review Criteria were provided by the Project Team—consisting of principal investigators, project managers, and others involved in the project. These criteria were slightly revised by the Technical Secretary and approved by the Project Team. The RP received documents describing various aspects of the project for their review. The summary of the project included in this report was prepared by the Technical Secretary using the same documents that had been provided to the RP. The Project Team received the project summary for review and approval. In addition, the staff of RSI undertook the task of preparing a list of acronyms. This list, as reviewed and approved by the Project Team, is also included in this report.

On August 5, 2004, during a telephone conference call, the RP was introduced to the American Society of Mechanical Engineers peer review process and the DOE's desire for a non-conflicted and independent peer review. Subsequently, individual members of the RP prepared findings and recommendations for specific review criteria, which were made available to the Technical Secretary of the RP who in turn prepared a draft of the findings. This draft was discussed among the members of the RP during a telephone conference on August 19, 2004. The recommendations resulted from an assessment of the entire findings by the RP. Therefore, a recommendation does not always follow a finding. Subsequent to the necessary copyediting, the report was finalized.

Consistent with the procedures established by the ASME/RSI process, the copyedited *Report of the Review Panel* was provided to DOE for identification of potential errors, misunderstandings, and areas of ambiguity. The Technical Secretary contacted the members of the RP reporting the comments received from DOE which were considered by the RP.

Charles O. Velzy
A. Alan Moghissi

Executive
Summary

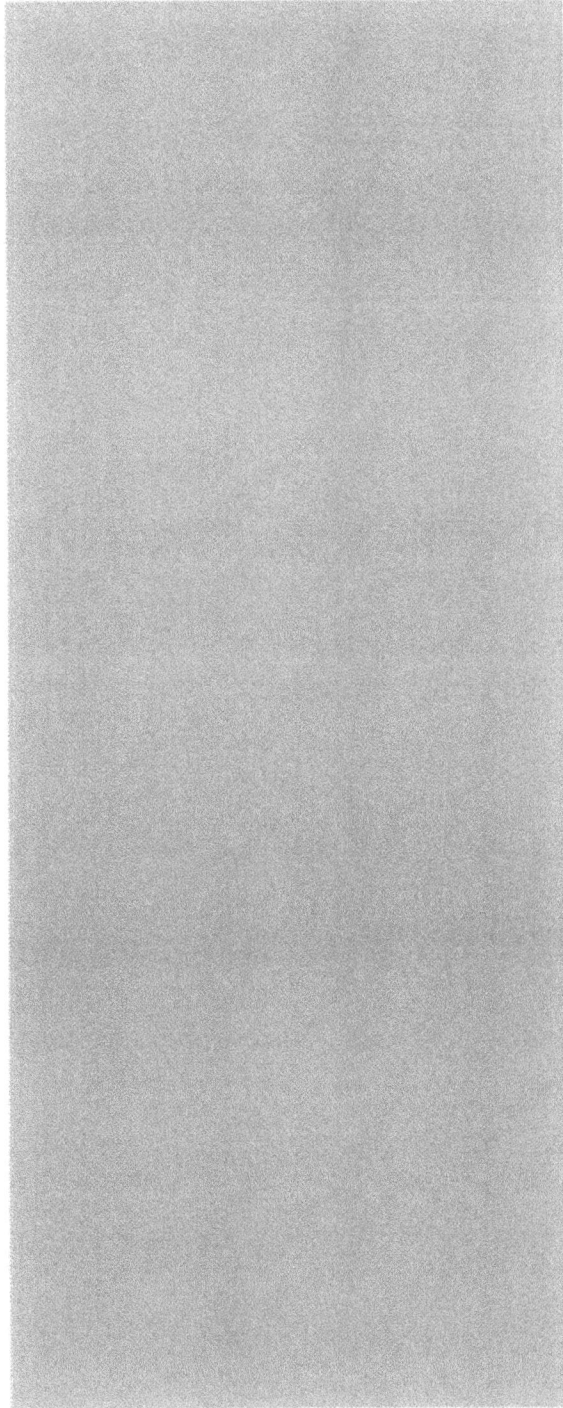

The objective of the Waste Package Closure System (WPCS) project is to assist in the disposal of spent nuclear fuel (SNF) and associated high-level wastes (HLW) at the Yucca Mountain site in Nevada. Materials will be transferred from the casks into a waste package (WP), sealed, and placed into the underground facility. The SNF/HLW transfer and closure operations will be performed in an aboveground facility.

The objective of the Control System is to bring together major components of the entire WPCS ensuring that unit operations correctly receive, and respond to, commands and requests for data. Integrated control systems will be provided to ensure that all operations can be performed remotely. Maintenance on equipment may be done using hands-on or remote methods, depending on complexity, exposure, and ease of access. Operating parameters and nondestructive examination results will be collected and stored as permanent electronic records. Minor weld repairs must be performed within the closure cell if the welds do not meet the inspection acceptance requirements. Any WP with extensive weld defects that require lids to be removed will be moved to the remediation facility for repair.

The Review Panel was presented with the following documents:

1. TFR-282 *Waste Package Closure System Technical Requirements Document*. This document outlines the technical requirements for the Waste Package Closure System addressing only the subsystems needed to perform waste package closure-related operations. Physical structures, utility needs, and connections from inside the structure surface are not within the scope of this document.
2. TFR-283 *Component Design Description: Welding and Inspection System*. This document presents design decisions for the overall configuration of the welding and inspection system. It also presents technical requirements for the overall system.
3. TFR-295 *Component Design Description: WPCS Safety System*. The objective of this document is to inform equipment designers of WPCS safety system requirements and capabilities so that they are aware of what the system offers for operating the equipment safely.
4. TFR-300 *Component Design Description: WPCS Control and Data Management System*. This Component Design Description document defines the design requirements and descriptions for the closure cell/operations gallery power and controls interface; support area/glovebox power and controls interface; operations gallery/support area controls interface; control electronics equipment locations; control software architecture; control software communications protocol to the hardware device control modules (HDCMs); software configuration management; database management; and DCMIS interface.
5. EDF-5103 *WPCS Welding Process: Control Functions and Associated Performance Requirements*. This document presents control functions needed for closure welding of Yucca Mountain waste packages by the Waste Package Closure System (WPCS) Welding and Inspection System. It also presents associated performance requirements for those control functions. EDF-5103 (INEEL 2004c) is a lower tier document to TFR-283 (INEEL 2004b).

The design of the control system is consistent with the derived technical requirements in TFR-282 *Waste Package Closure System Technical Requirements Document*.

The Waste Package Closure System (WPCS) will be located in the Yucca Mountain Project surface facility. It comprises all the structures and equipment located in the closure cells, closure-operating gallery, closure support area, closure maintenance areas, and maintenance area. The WPCS receives a waste package (WP) after it has been loaded with spent nuclear fuel/high level waste (SNF/HLW). It will be positioned below a process opening in the operating level floor of the closure cell. Although there will be several different WP lengths, the tops of the various WPs will be at the same height with respect to the top of the floor in the closure cells. The WP will be unshielded, but the inner lid will be in place. The spread ring assembly will not be inserted but will have been placed on the inner lid in the load cell before entering the closure area. The Technical and Functional Requirements (TFR)-282 (INEEL 2004a) document estimates the radiation field surrounding the WP about 1,500 rem/h (15 Sv/h) above the inner lid, and about 200 rem/h (2 Sv/h) to the side of the WP. The field will consist mainly of gamma radiation, but neutrons may also be emitted. Because of the high radiation field surrounding the WP, personnel will not be able to enter the closure cell when the WP is present in the cell. Thus, all closure operations and most maintenance in the closure cell will be performed remotely. The temperature is estimated to be about 200°F (93°C) on the inner lid.

The design of the control system is consistent with established scientific and engineering principles and standards. In particular, the PT has demonstrated through its design assumptions, design approach, and engineering analysis a thorough awareness of relevant published scientific and engineering information as well as practices of relevant industries.

For the prototype design considerations performed to date, the PT has clearly demonstrated thoroughness, good engineering principles, and awareness vis-à-vis safety, productivity, equipment costs, and reliability.

The PT has performed only a preliminary design. A prototype of the WPCS is currently being developed, constructed, and demonstrated. For purposes of the preliminary design completed, the PT has presented adequate technical documentation (such as functional and operational requirements; technical requirements; design analyses; and trade studies) to justify its preliminary design approach. However, details have been left open pending evaluation of the prototype under construction.

In its requirements and design concepts, the PT has adequately identified and addressed the occupational safety and health hazards related to the execution and operation of the control system. In particular, the PT has demonstrated adequate safety and health expertise as the system is designed, developed, and demonstrated. The PT has identified the need for radiation shielding windows; shield walls; HVAC systems to provide adequate air exchanges; remote maintenance areas to safely maintain and repair equipment; and to reduce risk of contamination spread and personnel exposure to radiation. The bounding and average dose rate information provided indicates that the PT has access to adequate safety and health expertise as the system is designed, developed, and demonstrated.

Based on a careful assessment of the information provided to the Review Panel (RP) and the findings developed in response to the review criteria, the RP provides the following recommendations:

1. This project should be continued. For the prototype design considerations performed to date, the PT has clearly demonstrated thoroughness, good engineering principles, and awareness vis-à-vis safety, productivity, equipment costs, and reliability.
2. The full and complete cable management system proposed by the PT should be implemented in the prototype currently under construction to allow thorough testing and evaluation of its expected longevity and overall reliability under the environmental conditions of the closure cell. In case the longevity is less than the 50-year specified lifetime, provisions should be made for replacement of the cable management system.
3. The PT should determine the impact of the pinion gear drive on the robotic arm performance. Additionally, the PT should consider the position resolution and repeatability of the robotic arm position controller, as well as the deflection characteristics of the robotic arm under load on the end-effector position resolution.

4. The PT should resolve the seeming discrepancy between the assumed positioning requirements of ±0.1 mm (±0.004 in.) and the stated minimum resolution of 0.016 in (0.4 mm) for the horizontal seam tracking system.
5. The PT should relate the stated automatic voltage control (AVC) resolution of ±0.05 V to estimated positioning accuracy in the vertical direction to ascertain whether the stated voltage control resolution is sufficient to meet the assumed vertical positioning accuracy requirements.
6. The PT should address methods of material removal and preparation for weld repairs; inspection of the prepped repair area; and repair welding in light of the substantially-greater demands on the robots' force and rigidity capabilities in performing the necessary grinding operations for defect removal.
7. The PT should develop a risk assessment to assist in the development of a test specification and test report for the safety control system.
8. The PT should provide all relevant details regarding the geometry and inventory of the waste containers packages expected to be received in the WPCS.

In addition, the RP provides the following recommendations that are not directly related to the Control and Data Management aspects:

1. As the WPCS design evolves from prototype to final design, the PT should thoroughly address and document the lifetimes and ability to operate the equipment including electronics (as in the case of video cameras), insulation, lubricants, and other materials expected in high radiation fields.
2. As the WPCS design evolves from prototype to final design, in order to maintain temperatures below the weld interpass maximum temperature of 350°F (177°C) for the 316 stainless steel components, and 200°F (93°C) for the Alloy 22 components, the PT should consider means of cooling the components between weld passes.
3. In the final design of the WPCS, the PT should consider and select a means of providing stress mitigation of the narrow-groove closure weld between the Alloy 22 outer lid and the Alloy 22 shell.
4. A suitable welder qualification program should be developed to ensure that weld operators can be relied upon to provide the very important wire guide positioning function in addition to their other duties at their remote location.
5. The proposed nondestructive examination (NDE) eddy-current and ultrasonic techniques should be tested and evaluated under operating conditions of radiation and temperature to determine the feasibility of the proposed approaches.
6. The PT should consider eddy-current testing for the middle lid prior to overall testing in order to evaluate the need for weld repair before attaching the upper lid.
7. The PT should evaluate in greater detail non-contact ultrasonic techniques such as laser ultrasonics; electromagnetic acoustic transducers (EMATs); and air-coupled ultrasonics for the inspection process.
8. The PT should develop a validation plan sufficient to qualify the selected NDE techniques for operation in a harsh environment with high reliability.
9. The PT should be aware of changing technologies as parts are replaced. The PT should determine the availability from vendors of the components used for the WP closure system, and either:
    9.1. develop a procurement plan for spare parts; or
    9.2. design provisions into the WP closure system for continuous upgrade of new components as they become available.
10. The PT should take into account the temperature differential between the lids and the WP in terms of fitup and welding.

Peer Review
Process

INTRODUCTION

There is consensus within the technical community on the definition, process, and key criteria for the acceptability of peer review. Peer review consists of a critical evaluation of a topic by individuals who—by virtue of their education, experience, and acquired knowledge—are qualified to be peers of an investigator engaged in a study. A peer is an individual who is able to perform the project, or the segment of the project that is being reviewed, with little or no additional training or learning.

Recognizing that peer review constitutes the core of acceptability of scientific and engineering information, virtually all professional societies of scientists and engineers have instituted formal procedures for peer review for their activities. The American Society of Mechanical Engineers (ASME), also known as ASME International, has over a century of experience in peer review. Consistent with its mission and tradition, ASME, in cooperation with the Institute for Regulatory Science (RSI), has established a peer review program devoted to the review of activities of various government agencies (ASME 2003, RSI 2003). The reports of the peer reviews resulting from this program have been published (ASME/RSI 1997, 1998, 1999, 2000, 2001a, 2001b, 2001c, 2002a, 2002b, 2002c, 2002d, 2002e, 2003a, 2003b, 2003c, 2004a, 2004b).

PEER REVIEW PROCESS

The structure of the peer review process established by the ASME/RSI team consists of a tiered system. For each specific area, the entire process is overseen by a committee. The review of specific topics is performed by Review Panels (RPs).

Peer Review Committee for Energy and the Environment

The Peer Review Committee for Energy and the Environment (PRCEE) is a standing committee of ASME formed to oversee peer review for one particular program in an agency. Its members are chosen on the basis of their education, experience, peer recognition, and contribution to their respective areas of competency. An attempt is made to ensure that all needed technical competencies and diversity of technical views are represented in the PRCEE. The members of the PRCEE must be approved by the Board on Research and Technology Development of the Council on Engineering of ASME. The PRCEE includes an Executive Panel (EP) that is responsible for the day-to-day operations of the PRCEE. Except for the EP, membership in ASME is not required for appointment to the PRCEE. As the overseer of the entire peer review process, the PRCEE enforces all relevant ASME policies, including compliance with professional and ethical requirements. A key function of the PRCEE is the approval of the appointment of members of RPs for a specific project.

Review Panels

The review of a project, a document, a technology, or a program is performed by an RP consisting of a small group of highly-knowledgeable individuals. Upon the completion of their task, the RPs are disbanded. The selection of reviewers is based on the competencies required for the specific review assignment. The number of individuals in an RP depends upon the complexity of the subject to be reviewed. The selection of a reviewer is based on the totality of that individual's qualifications. However, there are several generally-recognized and fundamental criteria for assessing qualifications of a reviewer. These are as follows:

1. Education: A minimum of a B.S. degree, preferably an advanced degree in an engineering or scientific field, is required for any peer reviewer.

2. Experience: In addition to education, the reviewer must have significant experience in the area that is being reviewed.

3. Peer recognition: Election to an office of a professional society, serving on technical committees of scholarly organizations, and similar activities are considered to be a demonstration of peer recognition.

4. Contributions to the profession: Contributions to the profession may be demonstrated by publications in peer-reviewed journals. In addition, patents, presentations at meetings where the papers were peer-reviewed, and similar activities are considered to be contributions to the profession.

5. Conflict of Interest: One of the most complex and contested issues in peer review is a set of subjects collectively called conflict of interest. The ideal reviewer is an individual who is intimately familiar with the subject and yet has no monetary interest in it. Despite this apparent difficulty, the ASME and similar organizations have successfully performed peer review without having a real or apparent conflict of interest. The guiding principle for conflict of interest is as follows: *Those who have a stake in the outcome of the review may not act as a reviewer or participate in the selection of reviewers.*

Due to the multidisciplinary nature of many projects reviewed by the ASME/RSI team, rapid identification of qualified peer reviewers and their availability to participate in the review process are key ingredients for a successful program. The process used for the identification of reviewers is multifaceted. The Administrative Manager of the Peer Review Program receives recommendations from sources within ASME; previous members of the RP; sister societies; other organizations and individuals; the U.S. Department of Energy (DOE); DOE contractors; and others. However, the selection of peer reviewers is based entirely on criteria identified by ASME. The details of various aspects of peer review, including conflict of interest, can be found in the ASME *Manual for Peer Review* (ASME 2003) and the associated procedures (RSI 2003).

COOPERATION WITH OTHER PROFESSIONAL SOCIETIES

The ASME is a large professional engineering society having in excess of 125,000 members. Although the predominant discipline of the members is mechanical engineering, there are members who—by virtue of their education, training, or experience—are competent in other disciplines. The Council on Engineering includes divisions ranging from classical mechanical engineering (design, heat transfer, and power) to solar engineering; environmental engineering; and safety and risk analysis. Despite the diverse competency within ASME, it is recognized that on occasion it will become necessary to peer review activities which include disciplines that are outside the areas of competency of ASME and its members. These disciplines may include geology, hydrology, toxicology, and ecology. Consequently, ASME has reached formal and informal agreements with its sister societies to identify qualified reviewers in areas outside of those covered by the membership of ASME.

PERFORMING ORGANIZATIONS

The Center for Research and Technology Development of ASME manages a number of scientific and engineering activities, including peer reviews. Because of ASME's conscious effort to maintain a small in-house staff, it relies upon other organizations to provide detailed project management services in its research, development, and similar activities. Accordingly, ASME and RSI joined forces in a collaborative effort to perform the peer review for the U.S. Department of Energy. While the ASME staff in Washington, DC provides the staff support for the PRCEE, the detailed management and staff support for the RPs is provided by RSI.

American Society of Mechanical Engineers

As one of the largest professional engineering societies, ASME has a long and distinguished history. Its activities are carried out primarily by members who volunteer their time in support of engineering and scientific advancement. For obvious reasons, ASME also has a paid staff to manage the day-to-day operations of such a large professional society. ASME has a detailed structure for its operation, consisting of councils, boards, divisions, and committees. The Council on Engineering has 38 divisions, including: Environmental Engineering; Solid Waste Processing; Nuclear Engineering; Safety Engineering; and Risk Analysis. The Council on Codes and Standards develops ASME codes and standards that are the backbone of many industries—including power production—worldwide. The Council on Codes and Standards is also responsible for the development of standards for activities such as certification of incinerator operators. The ASME was a founding member of the American Association of Engineering Societies and a founding member of the American National Standards Institute.

Institute for Regulatory Science

RSI is a not-for-profit organization chartered under section 501(c)3 of the Internal Revenue Service. It is dedicated to the idea that societal decisions must be based on the best available scientific and engineering information. According to the RSI mission statement, peer review is the foundation of the best available scientific and engineering information. Consequently, RSI has promoted peer review within government and industry as the single most important measure of reliability of scientific and engineering information. In its activities, RSI seeks the cooperation of scholarly organizations. Historically, a large number of RSI activities have been performed in cooperation with ASME. RSI is located in the Washington, DC metropolitan area.

Peer Review Criteria, Findings, and Recommendations of the Review Panel

PEER REVIEW CRITERIA AND FINDINGS OF THE RP

The Findings and Recommendations of the Review Panel (RP) are based on the independent peer review of requirements and capabilities in the following documents:

1. TFR-282 *Waste Package Closure System Technical Requirements Document* (INEEL 2004a) which presents the systems-level requirements.
2. The following four design-level documents:
   2.1. TFR-283 *Component Design Description: Welding and Inspection System* (INEEL 2004b)
   2.2. TFR-295 *Component Design Description: WPCS Safety System* (INEEL 2004d)
   2.3. TFR-300 *Component Design Description: WPCS Control and Data Management System* (INEEL 2004e)
   2.4. EDF-5103 *WPCS Welding Process: Control Functions and Associated Performance Requirements* (INEEL 2004c)

The findings of the RP with respect to the review criteria are as follows:

Criterion 1

Is the design of the control system consistent with the derived technical requirements in TFR-282 *Waste Package Closure System Technical Requirements Document*?

Finding 1 of the RP

The design of the control system is consistent with the derived technical requirements in TFR-282 *Waste Package Closure System Technical Requirements Document* (INEEL 2004a).

The Waste Package Closure System (WPCS) will be located in the Yucca Mountain Project surface facility. It comprises all the structures and equipment located in the closure cells, closure-operating gallery, closure support area, closure maintenance areas, and maintenance area. The WPCS receives a waste package (WP) after it has been loaded with spent nuclear fuel/high level waste (SNF/HLW). It will be positioned below a process opening in the operating level floor of the closure cell. Although there will be several different WP lengths, the tops of the various WPs will be at the same height with respect to the top of the floor in the closure cells. The WP will be unshielded, but the inner lid will be in place. The spread ring assembly will not be inserted but will have been placed on the inner lid in the load cell before entering the closure area. The Technical and Functional Requirements (TFR)-282 (INEEL 2004a) document estimates the radiation field surrounding the WP about 1,500 rem/h (15 Sv/h) above the inner lid, and about 200 rem/h (2 Sv/h) to the side of the WP. The field will consist mainly of gamma radiation, but neutrons may also be emitted. Because of the high radiation field surrounding the WP, personnel will not be able to enter the closure cell when the WP is present in the cell. Thus, all closure operations and most maintenance in the closure cell will be performed remotely. The temperature is estimated to be about 200°F (93°C) on the inner lid.

The WP consists of two containers, one inside the other. The inner vessel is made of stainless steel 316, and the outer shell is made of Alloy 22. The specific identity of the WP and all components of the WP entering the closure area will be verified before entry. Visual examination by video will be performed before welding to ensure cleanliness and to verify position. A stainless steel inner lid will be inserted into the inner vessel before the WP enters the WP closure area (WPCA). Component temperatures will be measured by a thermocouple before tack welding and before each weld pass. Weld interpass temperatures must not exceed 350°F (177°C) for the 316 stainless steel components, and 200°F (93°C) for the Alloy 22 components.

In the event these temperatures are exceeded, welding will not be performed, and suitable means will be taken to reduce the temperature of the components below the maximum allowed before welding. All welding will be performed using the cold-wire gas tungsten arc welding process. Since heat will be imparted to the parts during welding, it can be expected that the weld interpass temperatures will always exceed 200°F (93°C), and therefore part-cooling will be necessary. However, the cooling means are not considered or discussed in the documents presented to the RP. Moreover, the temperature differential between the lids and the WP has not been taken into account in terms of fitup and welding.

A one-segment spread ring will be used to mechanically retain the inner lid. The spread ring will be tack-welded into position, and the tack welds will be visually examined and dressed, as necessary. A two-pass seal weld will be made between the spread ring and the inner lid; spread ring and inner vessel; and spread ring segment ends. This seal weld will be visually inspected. The inner vessel will then be evacuated and backfilled with helium through a purge port in the inner lid. The vacuum will be held for not less than 30 minutes. The inner vessel will then be evacuated and backfilled a second time with helium without the 30-minute vacuum hold time. The purity of the helium will also be checked before backfilling. A leak test with a mass spectrometer will be performed to ensure that there is no helium in the region near the inner lid and associated seal welds. The purge port will then be plugged, leak tested, and covered by a purge port cap, which will be welded to the inner lid using a two-pass (minimum) seal weld. This seal weld will be visually inspected.

The Functional and Operational Requirement (F&OR) 1.1.2.3.2-3 of TFR-282 (INEEL 2004a) states that the amount of oxygen remaining in the WP shall be below predetermined limits (see Table 1, page 39 of this report). However, this limit is not provided.

The middle Alloy 22 lid will be tack-welded to the Alloy 22 shell and dressed as necessary followed by a multipass fillet weld. This weld will be visually and eddy-current inspected. A second Alloy 22 lid—the outer lid—will be placed in the Alloy 22 shell; tack-welded and dressed; and a multipass, full-thickness, narrow groove weld will be made joining the lid to the Alloy 22 shell. The method to achieve the spacing between the outer lid and the middle lid has not been specified. Ultrasonic inspection will be performed on the outer lid weld following each weld pass. The cover pass will be inspected using visual, eddy current, and ultrasonic methods. Following the inspection, the weld will be stress-mitigated by inducing residual compressive stresses in the upper surface of the weld. A second inspection using visual, eddy current, and ultrasonic methods will be performed after the stress mitigation.

The WPCS process has been designed to minimize the amount of time a WP stays in the closure cell in order to support the overall repository throughput goals. The closure process requires an estimated 44 hours to complete at 100% efficiency (no weld defects, operational delays, etc.). An estimated 60% efficiency (for maintenance, repairs, etc.) is assumed for the overall system, resulting in an estimated throughput time of up to 70 hours for all closure operations.

The Waste Package closure area will be maintained in a dry condition. Liquids are avoided in the WP closure cell to eliminate any chance of liquids entering a WP and to reduce additional waste streams. Liquid inside a WP would jeopardize the corrosion resistance of the package and could also provide the conditions for a criticality event. Normally, no free liquids will be introduced into the closure cell—an exception being small, controlled quantities of water used for ultrasonic inspection of the outer lid weld and the stress mitigation process. Preclusion of liquids is especially important before completion of the seal welds on the inner lid.

The WPCS includes all equipment and operations within the WPCA that directly support closure of a waste package. Ancillary facilities are specified and provided by the Facility design groups. The Facility design groups are responsible for moving the WP into the WPCA; securing it within the required tolerances; and moving the WP out of the WPCA once closure is completed. The Facility is also responsible for the moving door between the closure cell and the maintenance area.

The WPCS is located between the SNF/HLW transfer cell and the WP loadout cell. Several areas are used to complete the WPCS operations: the WP station on the ground floor; the closure cell, WP closure maintenance area, closure support area, closure operating gallery on the operating floor; and the maintenance area on the upper floor. Page 15 of TFR-282 (INEEL 2004a) shows viewing windows for the closure cell, but the viewing requirements are not specified, and the items that are to be reached by the master-slave manipulator are not identified (see Fig. 6, page 47 of this report). TR-3.1.1 of TFR-282 (INEEL 2004a) requires that the WPCS be designed to be compatible with the facility footprint (see Table 2, page 45 of this report). However, in order to ensure that the facility can accommodate the equipment and process, it is the facility footprint that needs to be compatible with the WPCS, not vice-versa.

Location of the WP in the WP station will be controlled by the Facility. It will be positioned within a 9-ft-(2.7-m)-diameter process opening: the center point 24 ft (7.2 m) from one side wall, 16 ft (4.8 m) from the opposite side wall, and 10 ft (3 m) from the operating gallery wall. The top of the WP must be within 7 ½ and 12 in (19 and 30 cm) below the top surface of the second floor, ±2 in (±5 cm) in the X and Y direction from the center of the process opening, and less than ±½ in (±1.25 cm) in flatness. These tolerances appear to be large when compared with the tolerances for locating end-effectors.

All operations on the WP (welding, non-destructive examination [NDE], inerting, stress mitigation) will occur in the closure cell. No access to the closure cell will be allowed when a WP is present, but a plate will be available to cover the process opening if needed during non-operating periods. Frequent maintenance activities will be performed in the glovebox, located in the closure support area. Weld tool trays will be moved from the closure cell into the glovebox by way of the transfer tunnel for change out of weld wire and calibration. The purge port tool and stress mitigation equipment may also be serviced in the glovebox. Transfer of materials (lids, etc.) into the closure cell will also be achieved using the transfer tunnel. Larger equipment (leak detection tool, master slave manipulators, cranes) can be routinely serviced in the WP closure maintenance area, at estimated intervals of 6 to 12 months.

Large process support equipment will be located in the maintenance area on the upper floor; for example, laser peening (if selected) for stress mitigation process and vacuum equipment. Hatches to the closure cell and closure support area allow the maintenance area crane to move equipment into the maintenance area for repair if necessary.

The TFR-282 (INEEL 2004a) requires that commercial equipment and products be used wherever practical to reduce cost and schedule and to improve maintainability of the overall WPCS. Commercial products are readily available and have shorter delivery times, known reliabilities, and available spare parts, and are generally lower in cost.

Individual components (such as welding or inerting) will have individual control features. An overarching Supervisory Control System (SCS) will also be employed to ensure that each individual operation is coordinated with the other operations and that conflicts are avoided. Some interface with the Digital Control and Management Information System (DCMIS) will be required. The closure cell shield door controls shall have an Ethernet/Transmission Control Protocol/Internet Protocol (TCP/IP) interface to the WPCS control system. An Ethernet/TCP/IP interface shall be provided at the closure cell control system for central control system communications. Communications between the various subsystems and the WP Closure Control System shall employ the Open Systems Interconnection Reference Model.

Heat generated by the contents of the waste package and by the welding operations must be considered in the operating environment analysis. Ionizing radiation from the waste package will be significant and is a major concern in the equipment design—particularly electronic items. Special considerations will be required in the design. Specific radiation fields will be estimated based on the fuel inventory and the cell design. Other cell environment issues, such as electromagnetic noise, will be addressed in the individual components as the design develops.

The waste package enters the closure cell loaded with SNF or HLW material. Although the inner lid will have been placed into position, there will be considerable radiation in the form of gamma radiation and some neutron radiation that will be emitted from the top of the waste package. This radiation is of sufficient intensity to pose a significant operational hazard to various components of the closure equipment, including electronics, insulation, lubricants, and other materials. Both their lifetimes and their ability to operate (in the case of video camera, for example) are issues that need to be addressed. As a general rule, although radiation-hardened components can be used in some cases, they are expensive. A solution is to either shield or locate such components out of the radiation. The highest radiation field is located directly over the top of the WP. Consequently, as many components as possible will be placed to the side of the WP in a location that can easily be shielded. In its preliminary prototype design considerations, the PT has not thoroughly addressed equipment lifetimes and the ability to operate in high radiation fields.

The technical requirements derived in TRF-282 call for stress mitigation of the narrow-groove closure weld between the Alloy 22 outer lid and the Alloy 22 shell. Laser peening is suggested in TRF-282 as one possible approach, and the maintenance area is sized to include large process support equipment such as the laser peening equipment. In its preliminary prototype design considerations, the PT has not addressed the required stress mitigation process.

The TFR-282 (INEEL 2004a) does not provide all the relevant dimensions of the waste packages as they are received from the waste generators. The PT mentions the different lengths for the incoming WPs, and the WP position tolerances relative to the hole. However, without knowing the diameters of the WPs and the diameters of holes, it is not possible to comment on the tolerances specified for the WPs.

In the Introduction of the TFR-282 document (INEEL-2004a), it is stated that the operations to be covered in the WPCS are for commercial and government-owned Spent Nuclear Fuel (SNF) and High Level Wastes (HLW). This needs clarification.

A commercial nuclear industry storage spent nuclear fuel canister is typically 5 to 6 ft. (1.5 to 1.8 m) in diameter while the DOE SNF canisters are approximately 1.5 to 2 ft (0.45 to 0.6 m) in diameter.

Further, if the report refers only to DOE SNF, among the many types of SNF in the DOE's inventory, the two major types of fuels to be identified for consideration are:

1. **Zirconium-Alloy clad fuels:** They are low-enrichment reactor fuels, and represent the largest number in the DOE SNF inventory with the greatest total mass and greatest mass of fissile and fertile heavy metals.

2. **Aluminum-alloy clad fuels:** They are high enrichment fuels used in Advanced Test Reactors (ATR). These fuels have the greatest burn-ups in the ATRs, and therefore contain some of the greatest fission product inventories of the fuels in large quantities.

TFR-282 (INEEL 2004a) refers to both SNF and HLW. However, in section 3.6.11 Radiation Hazards, the expected radiation dose rates are calculated only for 21-PWR fuels; no mention is made about the expected radiation dose rates for HLW. If HLW is included in the WPCS, implications of Fission Products and Actinides inventory; the container package details; and the container package integrity are relevant, and therefore need to be covered in the context of radiation hazard evaluations. Further, the source of the HLWs and their characteristics have to be identified: for example, whether the HLW is the vitrified product from the Savannah River plant or Hanford operations, these two having significantly different characteristics and nuclides inventories. Detailed information of the waste packages' contents is necessary for the maximum dose rate calculations, and adherence to the weld interpass temperature limits 350°F (177°C) (for the 316 stainless steel components) and 200°F (93°C) (for the Alloy 22 components).

TR-3.1-4 of TFR-282 (INEEL 2004a) states the equipment life is 50 years (see Table 2, page 45 of this report). However, based on the documents presented to the RP, the PT has not yet:

1. determined the availability from vendors of the components used for the WP closure system.
2. developed a procurement plan for spare parts or designed provisions into the WP closure system for continuous upgrade of new components as they become available.

The PT needs to be aware of changing technologies as parts are replaced.

Criterion 2

Is the design of the control system consistent with established scientific and engineering principles and standards? In particular, is the Project Team (PT) aware of the relevant published scientific and engineering information as well as practices of the relevant industry?

Finding 2 of the RP

The design of the control system is consistent with established scientific and engineering principles and standards. In particular, the PT has demonstrated through its design assumptions, design approach, and engineering analysis a thorough awareness of relevant published scientific and engineering information as well as practices of relevant industries.

The PT has chosen to weld the lids to the waste package vessels using the cold wire, gas tungsten arc welding process with tack welding used for initial setup. The basic configuration of the welding equipment will be a circular track machine. The circular track will be mounted a few inches above the operating floor of the closure cell, concentric to a large diameter hole in the operating floor, which will allow the waste package to be placed in position below the nominal center of the circular track for welding. Two carriages will be placed on the circular track to move two commercially-developed 6-axis robots around the waste package during welding. Control, data, power, and other utility cables (and hoses) will be run from the closure cell control area to these carriages. A cable management system will allow these cables to follow the carriage motion. The carriages will be capable of motions in excess of 180 degrees around the concentric track, which will allow for overlap of weld bead ends.

Section 2.1.2 of TFR-283 (INEEL 2004b) states that a pinion gear will be mounted to the output shaft of the motor, which will engage the ring gear mounted on the track. Gear backlash will impact the end effector position resolution. The documents presented to the RP do not provide information regarding this effect.

Section 2.1.3 of TFR-283 (INEEL 2004b) indicates that 6-axis or 7-axis robotic arms will be used for welding, quantitative and qualitative inspection of the welds. The documents presented to the RP do not provide information regarding the position resolution and repeatability of the position controllers for the robotic arms. The deflection characteristics of the robotic arms were also not presented. Moreover, 7-axis robotic arms are not yet commercially available.

Two identical welding end-effectors (one per robot) will be used for welding both the 316 stainless steel and Alloy 22 components. Each welding end-effector will incorporate means for quantitative visual inspection of the weld passes, as required. These inspections will be performed using a seam-tracking sensor. This sensor, when scanned along the weld joint, will be capable of making 3-dimensional measurements of both the weld joint and weld surface profiles, and will be mounted ahead of the welding torch.

Each end-effector will incorporate a weld vision camera that will have suitable arc light attenuation capabilities to provide video images of the weld pool during welding. The camera will be mounted in front of and behind the welding torch. Each welding end-effector will also incorporate video cameras to obtain video images of the weld-bead leading edge and behind the welding torch.

The end-effectors will incorporate a remotely-adjustable filler wire guide mechanism to ensure that the filler wire enters the weld pool in such a manner as to facilitate making a good weld. Each welding end-effector will incorporate a thermocouple or other temperature sensor capable of measuring the temperature of the base metal before each weld pass.

The welding end-effectors will incorporate a mechanism to provide lateral motion of the weld torch normal to the welding direction. This mechanism will be used for seam tracking and weld torch oscillation during welding. Each welding end-effector will also incorporate a mechanism to provide vertical motion of the torch parallel to the axis of the torch. This mechanism will be used for automatic voltage control during welding and arc touch-starting.

With this robot manipulator/end-effector approach described in TFR-283 (INEEL 2004b), the robot is depended on for relatively gross motion accuracy (± 1 mm or ± 0.04 in.) with the horizontal and vertical seam tracking systems providing the precision positioning (± 0.1 mm or ± 0.004 in.) required for welding. The PT opines that this approach is superior to one using the robot arm alone for positioning for the following reasons (INEEL 2004b):

1. Seam tracking and automatic voltage control are the highest bandwidth motion control functions required for closure of a waste package.
2. Relative high bandwidth motion of small hardware components (e.g., 5 lb $\approx$ 2.2 kg) is easier and more reliable than similar motion of large hardware components (e.g., 500 lb $\approx$ 226 kg).
3. Simultaneous seam tracking, torch oscillation, automatic voltage control, and arc starting are at the leading edge of robotic technology today.

The PT has chosen to use a cable management system to control the motion of hoses and electrical cables running between the closure cell wall and the carriages. A cable management chain will support the various cables and hoses, operating in a semi-circular tray outside the circumferential track. While the PT considered several alternative approaches to the cable management problem, it appears that there are concerns regarding the longevity and overall reliability of the cable management system. In case the longevity is less than the 50-year specified lifetime, it is necessary to make provisions for replacement of the cable management system.

The PT proposes to use commercial off-the-shelf control system components to monitor and control the various functions of the welding system. A human/machine interface (HMI) will allow human operators to operate and interact with the welding high-level motion trajectory planning and process controls system as an integral part of the outer loop control structure. The HMI will incorporate a graphical user interface (GUI) displayed on one or more component screens, with additional control devices, such as joystick, mouse, track ball, touch screen, inter-operator voice communications, within the closure cell operator workstations. Manual operator trajectory and process control offsets will be implemented through the HMI and GUI interfaces as part of the outer loop trajectory guidance algorithms. These operator offsets are transmitted as set-point or baseline trajectory offsets onto lower level set-point and tracking controllers operated within the welding control subsystems.

The welding system control architecture will be segmented into two primary levels:

1. Low-level set-point and trajectory-based trackers, implemented within subsystem control components or within the weld system control computers.
2. Higher-level motion trajectory planning and process controls in which the welding operators will be an integral part of the control loop design, implemented within the closure cell operator station computer system.

Moreover, subsystem controllers will be utilized to the greatest intent practicable, while higher-level control functions will be provided in the operator interface and supervisory control system functions.

The filler wire position control will comprise two parts:

1. The high-level weldstation operator-controlled GUI for setting and maintaining outer-loop set-points
2. The low-level hardware/software regulator module

The weldstation operator will be presented a live image of the weld wire with respect to weld pool, so that the weld wire position can be maintained in an optimal welding position based on user experience and existing welding procedures. In this case, the weldstation operator acts as the outer-loop controller for wire positioning. Constant monitoring of the weld-wire-to-weld-pool relationship will be needed due to its coupling with the automatic voltage control (AVC) algorithm.

The experience of a welder controlling the wire guide position when stationed directly at the equipment does not necessarily translate to performing the same function remotely with the aid of camera-generated images of a small portion of the arc area. A suitable welder qualification program needs to be developed to ensure that weld operators can be relied upon to provide the important wire-guide positioning function in addition to their other duties at their remote location.

Section 2 TFR-295 (INEEL 2004d) describes the safety system as comprising three processors (manufactured by three different vendors) designed to work in concert on the principle of a triple voting system. However, it is unclear whether the three processors perform different functions or the same function—thus providing redundancy. Appropriate evaluation laboratories (e.g., Underwriters Laboratories, Canadian Standard Association) will certify the triple voting system. These testing agencies establish testing standards for devices to insure safety from risk of fire, electric shock, or injury. Accordingly, they will recognize and list equipment, and will list the integration and the installation of equipment. However, the operational performance of the triple voting system must be validated through the development of a test specification, and documented in a test report. The PT did not present information regarding a risk assessment of the safety system. A risk assessment is necessary to assist in the development of a test specification and test report for the safety control system. EN 292-1 (ESO 1992a), EN 292-2 (ESO 1992b), EN 418 (ESO 1993a), EN 775 (ESO 1993b), EN 954-1 (ESO 1997a), and EN 1050 (ESO 1997b) are the normative documents to be used in the development of the risk assessment.

Design Requirement SS-6 of TFR-295 (INEEL 2004d) indicates that the safety system software will be verified through observation without specifying any guidelines; verification through testing is more appropriate (see Table 23, page 81 of this report). Similarly, as shown in Table 23, Design Requirement SS-11 of TFR-295 (INEEL 2004d) indicates that the safety system power will be verified through observation; verification through measurement is feasible and more appropriate.

Section 2.1 of TFR-295 (INEEL 2004d) mentions the "master/slave manipulator lockup" interface without clarifying whether the master/slave manipulators are operated manually or through the control system.

On page 4 of TFR-283 (INEEL 2004b) it is indicated that the "subsystems will be modular" without stating how this is to be accomplished. Section 1.2.2 Motion Control of TFR-283 (INEEL 2004b) does not state how the outer lid will be spaced from the middle lid prior to performing the vee groove weld. Section 2.1.7 of TFR-283 (INEEL 2004b) does not consider eddy-current testing for the middle lid prior to overall testing; this testing can evaluate the need for weld repair before attaching the upper lid.

Section 8.2.4.7 of EDF-5103 (INEEL 2004c) states that the "trajectory planner will search for a weld joint location within a predetermined location and area on the waste package, coordinated via general waste package coordinates provided by the closure cell supervisory control system. The planner searches using the seam-tracking sensor and weld end-effector cameras assisted by the weld station operator via the GUI system. Once the nominal weld joint location has been determined within the robot's coordinate frame, the robot and carriage baseline trajectories will be generated based on expected waste package geometry." An alternative for developing the manipulator arm trajectory is based upon a known reference position and orientation for the WP. This alternative requires the cell floor opening to be circular and lined with a machined ring. This machined ring becomes a reference point for determining both the center point and the orientation of the WP. Incorporating a 3D inspection system into the robotic manipulator, the slope of the WP relative to the cell floor can be determined by touching three points on the top of the machined ring to establish a plane; and then three points on the top of the WP. The position of the WP relative to the cell floor opening can be determined by touching four points around the circumference of the machined ring inside diameter and then touching four points around the circumference of the WP outside diameter. The position and orientation of the WP is known relative to a fixed position and orientation. This information can be used in conjunction with the seam-tracking information, or it can be used in lieu of the seam-tracking information.

For the prototype design considerations performed to date, the PT has clearly demonstrated thoroughness, good engineering principles, and awareness vis-à-vis safety, productivity, equipment costs, and reliability.

Criterion 3

Has the PT presented adequate technical documentation (such as functional and operational requirements; technical requirements; design analyses; and trade studies) to justify its design approach?

Finding 3 of the RP

The Review Panel has been presented with the following documents:

1. TFR-282 *Waste Package Closure System Technical Requirements Document* (INEEL 2004a). This document outlines the technical requirements for the Waste Package Closure System addressing only the subsystems needed to perform waste package closure-related operations. Physical structures, utility needs, and connections from inside the structure surface are not within the scope of this document.
2. TFR-283 *Component Design Description: Welding and Inspection System* (INEEL 2004b). This document presents design decisions for the overall configuration of the welding and inspection system. It also presents technical requirements for the overall system.
3. TFR-295 *Component Design Description: WPCS Safety System* (INEEL 2004d). The objective of this document is to inform equipment designers of WPCS safety system capabilities so that they are aware of what the system offers for operating the equipment safely.
4. TFR-300 *Component Design Description: WPCS Control and Data Management System* (INEEL 2004e). This Component Design Description document defines the design requirements and descriptions for the closure cell/operations gallery power and controls interface; support area/glovebox power and controls interface; operations gallery/support area controls interface; control electronics equipment locations; control software architecture; control software communications protocol to the hardware device control modules (HDCMs); software configuration management; database management; and DCMIS interface.
5. EDF-5103 *WPCS Welding Process: Control Functions and Associated Performance Requirements* (INEEL 2004c). This document presents control functions needed for closure welding of Yucca Mountain waste packages by the Waste Package Closure System (WPCS) Welding and Inspection System. It also presents associated performance requirements for those control functions. EDF-5103 (INEEL 2004c) is a lower tier document to TFR-283 (INEEL 2004b).

The PT has performed only a preliminary design. A prototype of the WPCS is currently being developed, constructed, and demonstrated. For purposes of the preliminary design completed, the PT has presented adequate technical documentation (such as functional and operational requirements; technical requirements; design analyses; and trade studies) to justify its preliminary design approach. However, details have been left open pending evaluation of the prototype under construction.

TR-3.4-2 of TFR-282 (INEEL 2004a) uses the word "preferred" with respect to failure modes (see Table 3). In fact there is no "preferred" failure mode. "Probable" failure mode is a better wording. The definition for "layup", as provided on page 33 of TFR-282 (INEEL 2004a), is incomplete, because it does not specify that the facility (or some system/subsystem) is not operating. Moreover, the definitions for subsystem, system, and WPCS from TFR-282 (INEEL 2004a) are not consistent with the corresponding definitions from TFR-295 (INEEL 2004d). It is not clear why there are markings on the left margin of pages vii and 1-10 of TFR-295 (INEEL 2004d). If these markings are indicators of revisions, then it is necessary to delete them.

The Introduction of TFR-283 (INEEL 2004b) states that this document presents "design decisions;" however, based on the nature of this document, "design descriptions" is more appropriate. On page 2 of

27

TFR-283 (INEEL 2004b), there is a reference to Subsection IX of the ASME (2001) code; in fact, this is Section IX covering non-destructive examination and welding repair. Moreover, it is unclear how visual inspection can be quantitative as mentioned at the bottom of the same page 2 of TFR-283 (INEEL 2004b). Item 3 on page 4 of TFR-283 (INEEL 2004b) mentions that "custom hardware may be used when neither of the above two options are adequate;" in this case "available" is the proper qualifier instead of "adequate."

Page 8 of TFR-300 (INEEL 2004e) clearly illustrates the display seen at the operator workstation, but it does not indicate the controls or control location associated with this station (see Fig. 18, page 90 of this report). EDF-5103 (INEEL 2004c) has no drawings illustrating the weld torch positions.

Based on prior experience, the PT assumes the welding torches need to be positioned laterally and vertically within about ±0.1 mm (±0.004 in) of the desired position over the weld joint during welding. This level of assumed accuracy is consistent with high quality gas tungsten arc welding. The PT's design approach is to employ two nominally-identical 6-axis robotic arms mounted to each of two carriages with each robot manipulator equipped with a welding end-effector with seam-tracking capability in both horizontal and vertical directions with respect to the weld joint. The positioning capabilities of the robotic manipulators in positioning the torch with respect to the weld joint is assumed to be ±1 mm (±0.04 in.). The PT's design approach relies on the positioning resolution of the horizontal and vertical seam-tracking systems to achieve the overall positioning accuracy of ±0.1 mm (±0.004 in.).

The PT proposes to use a commercial grade seam tracker for precise tracking of the joint in the horizontal direction. The unit will return geometric information pertaining to the measured weld joint centerline, as well as overall measured geometric weld joint parameters (e.g., weld joint outline). It will present the information in such a form that weld joint geometry might be evaluated (e.g., width, depth). There will be provision for display of such geometric parameters sent both to the weld station operator (via HMI) and to GUI applications. Furthermore, the unit will provide feedback on the weld geometry prediction (i.e., if the seam-tracking algorithm cannot determine the weld joint during any frame, it will report this fact). The PT states that the seam tracker used for these purposes will have a minimum resolution of 0.016 inches (0.4 mm). This resolution is in contradiction with the overall assumed positioning requirements of ±0.1 mm (±0.004 in.).

The PT proposes to use a servomechanism to provide vertical motion of the weld torch parallel to the axis of the torch. The feedback signal for this control system will be proportional to the sensed arc voltage. By comparing the sensed arc voltage with a desired set point reference, the servomechanism moves the torch in or out in the vertical direction as required to maintain the arc voltage at the desired set point. This type of servo is referred to in the industry as an Automatic Voltage Control (AVC) system. Since the relationship between arc voltage and arc length is nominally linear, the AVC serves as an accurate means of maintaining desired torch position in the vertical direction. The PT states that the AVC will have a voltage control resolution of ±0.05 V, and a controlled voltage range of 8 to 20 V. The PT does not appear to relate the AVC resolution in voltage to positioning accuracy in the vertical direction to ascertain whether the stated voltage control resolution is sufficient to meet the assumed vertical positioning accuracy requirements.

While the slope of the arc voltage versus arc length relationship is nominally linear, the slope varies significantly as a function of welding current. Since this slope directly affects the overall closed-loop gain of the AVC servo system, stability issues may arise over widely-varying currents, as typically experienced in the start and stop portions of the weld cycle. In some cases, this potential problem may be alleviated by deactivating the AVC during the upslope and downslope parts of the weld. In other more critical cases, however, vertical positioning control is desired over the full downslope period to avoid crater cracks in the weld termination. In this case, more involved adaptive control approaches may be required in the AVC implementation.

The TFR-282 (INEEL 2004a) calls for the ability to repair weld defects found by the NDE inspection systems. In the event of weld repairs, the weld volume containing the defect will have to be ground out, prepped for re-welding, and welded. The documents provided to the RP do not present any considerations of these repair operations in the robotic approach chosen, and do not address potential problems associated with material removal and preparation of the repair area; inspection of the prepped area; and repair welding. Use of the robotic manipulators for repair grinding operations is likely to place substantially greater demands on the robots' force and rigidity capabilities.

Conventional welding and inspection methods cannot be applied in this case of WP closure due to the harsh environment and the lack of shielding. Helium leak testing, remote welding, and real time NDE are important. In addressing these issues, qualified data must be generated to provide reasonable assurance that the fabrication process will produce a WP with high-quality welds made in accordance with welding procedures qualified to American Society of Mechanical Engineers (ASME 2001). Welding procedures, NDE procedures, and personnel must be qualified to the requirements set forth in this code. Joint designs that minimize welding defects; minimize weld process heat input; and facilitate inspection (especially volumetric) are required. All final welds must be inspected using visual, surface, and volumetric examination techniques. Remotely-operated equipment and techniques must be developed to perform closure weld inspections in the hot cell environment.

The PT has provided a description of a suggested set of NDE testing procedures, but has not evaluated their performance in a harsh environment. Both the proposed eddy current and phased array ultrasonic testing techniques need to be evaluated under operating conditions for signal/noise ratio, repeatability, and durability (transducers and cabling). The eddy-current probe needs to be evaluated for wear. Moreover, it is necessary to assess whether probe wobble, mechanical compliance, or liftoff (due to weld roughness or imprecise alignment) are problems. The ultrasonic-phased array probe needs to be evaluated for:

1. loss of signal due to water couplant problems and/or possible damage to the piezoelectric properties.
2. ability to scan the entire weld walls as well as the weld volume.

Other possibilities, such as non-contact ultrasonic techniques, may be helpful in case the water couplant problem for the ultrasonic technique is insurmountable due to temperature problems.

The documents provided to the RP do not present neither validation procedures for the NDE testing nor a comprehensive validation. The validation must include:

1. representative weld samples prepared with a statistically-significant number of well-characterized defects
2. scan plan designed to improve inspectability; maximize the probability of flaw detection; and minimize the number of false calls.

The documents presented to the RP do not mention any monitoring for surface contamination of the waste packages as they are accepted for closure in the WPCS.

Criterion 4

Have the occupational safety and health hazards related to the execution and operation of the control system been adequately identified and addressed in the requirements and design concepts of the control system? In particular, does the PT have access to adequate safety and health expertise as the system is designed, developed, and demonstrated?

Finding 4 of the RP

In its requirements and design concepts, the PT has adequately identified and addressed the occupational safety and health hazards related to the execution and operation of the control system. In particular, the PT has demonstrated adequate safety and health expertise as the system is designed, developed, and demonstrated. The PT has identified the need for radiation shielding windows; shield walls; HVAC systems to provide adequate air exchanges; remote maintenance areas to safely maintain and repair equipment; and to reduce risk of contamination spread and personnel exposure to radiation. The bounding and average dose rate information provided indicates that the PT has access to adequate safety and health expertise as the system is designed, developed, and demonstrated.

The safety system for the WPCS is designed to protect equipment from damage and personnel from injury. The WPCS safety system features a class of programmable logic controllers (PLCs) specifically designed for use in safety-critical applications. These modular units incorporate multiple independent processors that separately monitor inputs to determine their validity. If a trip occurs, these processors automatically compare data to determine whether the trip is valid. This tends to eliminate nuisance trips or false alarms that occur with safety relay systems.

Programmable safety systems also can monitor input switch/relay circuits and cabling, and can be programmed to alarm if the circuit opens or shorts. All programmable safety PLCs provide emergency stops, enabling devices, and safeguard or protective devices, in accordance with International Standards Organization and American National Standards Institute standards.

The WPCS safety system can be programmed to protect specific machines, zones, or areas, and can be programmed to output either to a specific piece of equipment or to the whole system. The system can be designed to monitor continuity through alarm switches and interlocks to help ensure that the systems are in order. The safety system interfaces with equipment throughout the closure cell area. The following list identifies specific equipment interfaces that have been defined by the PT: supervisory control system; remote handling system; welding systems; closure cell crane; master/slave manipulator lockup; and glovebox interlock controls. The safety system will be located in the equipment racks in the operating gallery. The architecture is a modular design that can be easily expanded. A standard software protocol will be used to communicate with the supervisory control system through Ethernet (TCP/IP).

The safety system will maintain all interlocks, emergency shutdown capabilities, and a network connection to the supervisory control system. If an interlock fails or an emergency stop occurs, the safety system will prevent operation of the equipment (safe stand down and lock) and report the occurrence to the supervisory control system (which has the capability of communicating to the DCMIS). Once a shutdown has occurred, an operator will be required to follow a startup procedure to bring the equipment back on line. Inherent to the safety system is constant monitoring and verification of the interlock/emergency switch circuits. Password authorization is required to modify the safety system software. However, it is unclear whether the validation of the software is part of the safety system modification process. A separate computer with an RS-485 interface capability and the appropriate development software is required to develop or modify the program. The Ethernet/TCP/IP interface does not allow program access.

The remote welding, inspection, and repair approach presented by the PT significantly reduces personnel radiation exposure; reduces secondary waste; and improves productivity. The modular integrated system

30

for automated welding significantly reduces personnel exposure during operation by providing one complete system and one procedure. The NDE technology normally available from commercial vendors is for post-weld testing that requires additional setup and testing procedures independent of the welding process. In addition to the staging, setup, and removal of the automated welding equipment, technicians and quality inspectors are required to setup and then remove the NDE equipment. These extra steps can be eliminated, thus significantly reducing the exposure time with the integration of the in situ NDE technology proposed by the PT. Additional exposure associated with post-weld repairs would be reduced by providing notification of weld defects on partially-completed welds, thus reducing the amount of weld filler material that is normally removed by grinding on completed welds.

The PT has presented best practices in hot cell design to minimize environmental/human health risks. The minimization of the secondary waste streams was adequately addressed. For example, the secondary waste stream produced through grinding is minimized by:

1. performing pass-by-pass inspection.
2. producing a smooth cap pass to enable direct inspection.

This implies reduced airborne contamination and less time associated with weld repair.

Page 15 of TFR-282 (INEEL 2004a) shows a HEPA filter (see Fig. 6, page 47 of this report), but the PT provides no criteria regarding these filters including their change-out. Moreover, there are no specifications regarding the temperatures to be maintained in the various areas, like the closure-operating gallery where operators are located and thus may require air conditioning.

RECOMMENDATIONS

Based on a careful assessment of the information provided to the Review Panel (RP) and the findings developed in response to the review criteria, the RP provides the following recommendations:

1. This project should be continued. For the prototype design considerations performed to date, the PT has clearly demonstrated thoroughness, good engineering principles, and awareness vis-à-vis safety, productivity, equipment costs, and reliability.
2. The full and complete cable management system proposed by the PT should be implemented in the prototype currently under construction to allow thorough testing and evaluation of its expected longevity and overall reliability under the environmental conditions of the closure cell. In case the longevity is less than the 50-year specified lifetime, provisions should be made for replacement of the cable management system.
3. The PT should determine the impact of the pinion gear drive on the robotic arm performance. Additionally, the PT should consider the position resolution and repeatability of the robotic arm position controller, as well as the deflection characteristics of the robotic arm under load on the end-effector position resolution.
4. The PT should resolve the seeming discrepancy between the assumed positioning requirements of ±0.1 mm (±0.004 in.) and the stated minimum resolution of 0.016 in (0.4 mm) for the horizontal seam tracking system.
5. The PT should relate the stated automatic voltage control (AVC) resolution of ±0.05 V to estimated positioning accuracy in the vertical direction to ascertain whether the stated voltage control resolution is sufficient to meet the assumed vertical positioning accuracy requirements.

6. The PT should address methods of material removal and preparation for weld repairs; inspection of the prepped repair area; and repair welding in light of the substantially-greater demands on the robots' force and rigidity capabilities in performing the necessary grinding operations for defect removal.
7. The PT should develop a risk assessment to assist in the development of a test specification and test report for the safety control system.
8. The PT should provide all relevant details regarding the geometry and inventory of the waste containers packages expected to be received in the WPCS.

In addition, the RP provides the following recommendations that are not directly related to the Control and Data Management aspects:

1. As the WPCS design evolves from prototype to final design, the PT should thoroughly address and document the lifetimes and ability to operate the equipment including electronics (as in the case of video cameras), insulation, lubricants, and other materials expected in high radiation fields.
2. As the WPCS design evolves from prototype to final design, in order to maintain temperatures below the weld interpass maximum temperature of 350°F (177°C) for the 316 stainless steel components, and 200°F (93°C) for the Alloy 22 components, the PT should consider means of cooling the components between weld passes.
3. In the final design of the WPCS, the PT should consider and select a means of providing stress mitigation of the narrow-groove closure weld between the Alloy 22 outer lid and the Alloy 22 shell.
4. A suitable welder qualification program should be developed to ensure that weld operators can be relied upon to provide the very important wire guide positioning function in addition to their other duties at their remote location.
5. The proposed nondestructive examination (NDE) eddy-current and ultrasonic techniques should be tested and evaluated under operating conditions of radiation and temperature to determine the feasibility of the proposed approaches.
6. The PT should consider eddy-current testing for the middle lid prior to overall testing in order to evaluate the need for weld repair before attaching the upper lid.
7. The PT should evaluate in greater detail non-contact ultrasonic techniques such as laser ultrasonics; electromagnetic acoustic transducers (EMATs); and air-coupled ultrasonics for the inspection process.
8. The PT should develop a validation plan sufficient to qualify the selected NDE techniques for operation in a harsh environment with high reliability.
9. The PT should be aware of changing technologies as parts are replaced. The PT should determine the availability from vendors of the components used for the WP closure system, and either:
   9.1. develop a procurement plan for spare parts; or
   9.2. design provisions into the WP closure system for continuous upgrade of new components as they become available.
10. The PT should take into account the temperature differential between the lids and the WP in terms of fitup and welding.

Project
Summary

INTRODUCTION

Nuclear reactors have been operating for over 50 years in the United States, and in many other countries in the world. They are used for electricity generation, research, and production of specific radionuclides. Once the fissionable material in the reactor fuel has been consumed, the spent nuclear fuel (SNF) is removed and is considered available for reprocessing or long-term storage.

Yucca Mountain Nevada is designated as the proposed geological repository for disposal of SNF and HLW. The U.S. Department of Energy (DOE) is preparing a license application to be submitted to the U.S. Nuclear Regulatory Commission (USNRC).

During repository operations, commercial and government-owned SNF and HLW will be loaded into casks and shipped to the Yucca Mountain repository, where it will be transferred from the casks into waste packages, sealed, and placed into the underground facility. Transfer of the SNF, HLW, and closure operations will be performed in a facility above the surface. Closure operations include sealing the waste package and all associated functions, such as welding the lids onto the waste package; filling the inner container with an inert gas; performing a nondestructive examination of the welds; and conducting stress mitigation. The Waste Package Closure System (WPCS) encompasses all of these operations.

The WPCS project, as summarized in this report, addresses control-related subsystems needed to perform waste package closure-related operations. It includes:

1. Technical requirements for the WPCS
2. Component design descriptions for the Welding and Inspection System; Safety System; and Control and Data Management System
3. Control functions and associated performance requirements for the Welding Process

There are numerous other requirements (including walls, floors, utility needs, and connection from inside the structure surface) that are not considered here as they are addressed by facility design groups.

CONTROL SYSTEM TECHNICAL REQUIREMENTS

WASTE PACKAGE CLOSURE SYSTEM DESCRIPTION

The following material is derived from the TFR-282, Waste Package Closure System Technical Requirement (INEEL 2004a) document.

The Yucca Mountain Project Repository Facility will house the Waste Package Closure System (WPCS). The WPCS will comprise five major areas in the Facility, as follows:

1. One or more closure cells
2. One closure support area
3. One closure operating gallery
4. One or more waste package closure maintenance cells
5. One maintenance area above the closure cell and closure support area

Figure 1 shows the WPCS layout for the Facility.

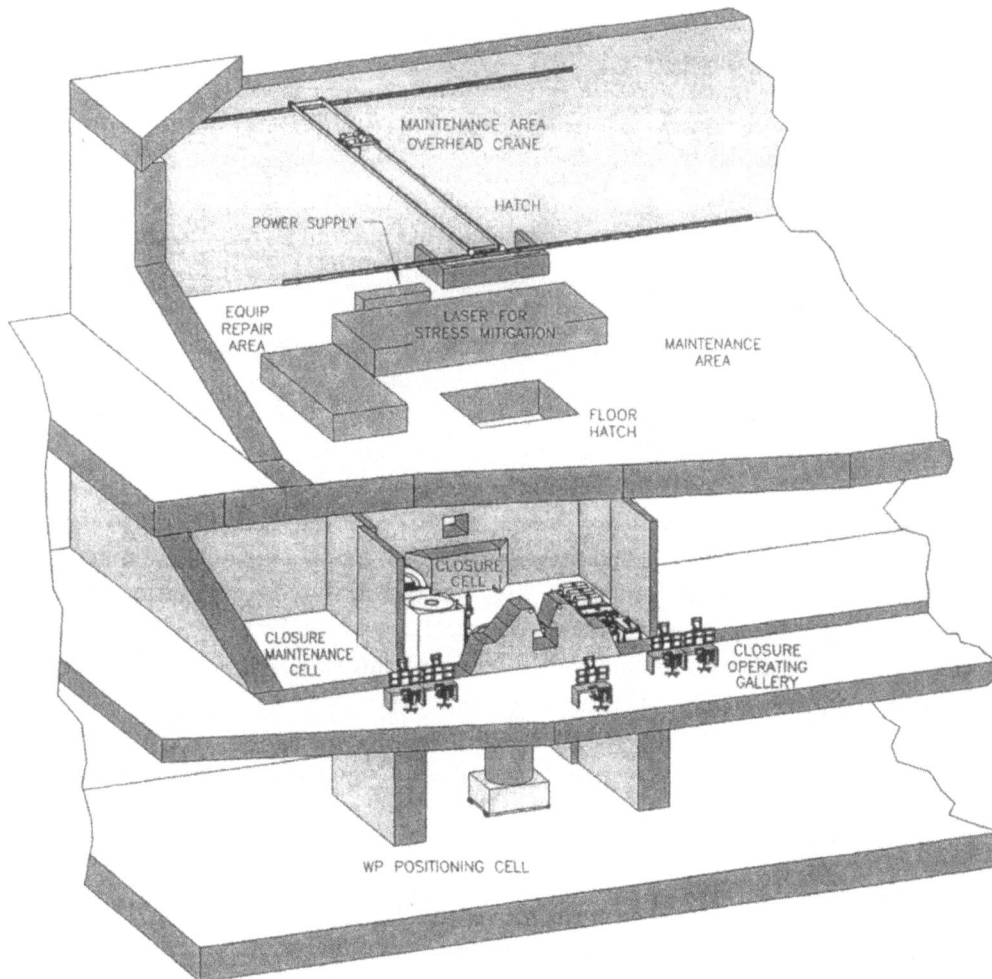

Fig. 1. Waste Package Closure System layout for the Yucca Mountain Facility (INEEL 2004a).

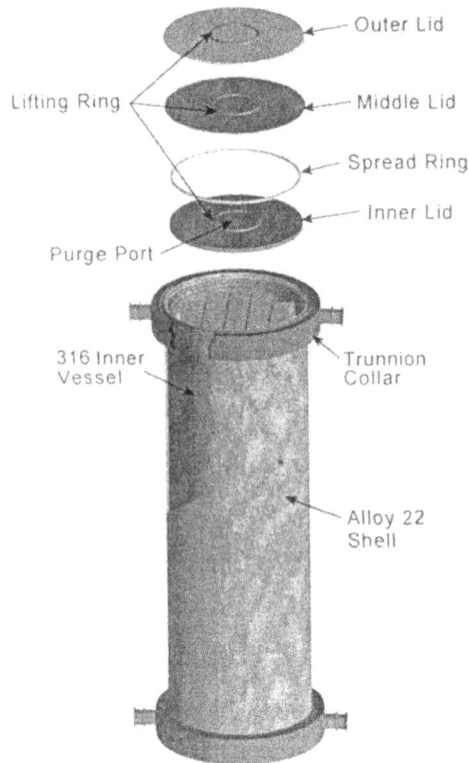

Fig. 2. The waste package (INEEL 2004a).

Closure of the waste package includes multiple operations that must all be performed remotely. Three lids must be welded onto the waste package; an inner lid is welded onto a stainless steel inner vessel; and a middle and an outer lid are welded to the Alloy 22 shell (Fig. 2). Visual inspections before welding are necessary to ensure cleanliness. Nondestructive examination of the final welds is required to verify weld integrity. In-process inspections and repairs on the outer lid are required to improve the productivity and efficiency of the closure system. A stress mitigation system will be employed to reduce the risk of stress-related failures because of the internal stresses induced in the weld area from the welding operations.

The inner vessel will be filled with an inert gas; sealed; and leak checked, in order to: 1) verify that the inner vessel is an inert environment that will reduce the risk of internal corrosion; and 2) facilitate heat transfer in the underground emplacement. A cap will be welded over the port used for inserting the inert gas. This weld will be nondestructively examined. Integrated control systems will ensure that all operations can be performed remotely. Maintenance on equipment may be done using hands-on or remote methods, depending on complexity and frequency. Operating parameters and nondestructive examination results will be collected and stored as permanent electronic records. Finally, minor weld repairs must be performed within the closure cell if the welds do not meet the inspection acceptance requirements. Waste packages with extensive weld defects that require a lid to be removed will be moved to the remediation facility for repair. The remediation system is outside the scope of the WPCS.

Functional and operational requirements allocated to WPCS

The facility-wide *Project Functional and Operational Requirements* document (Bechtel SAIC 2004) allocates one overall function ("seal WP") to the WPCS. It further decomposes the single function into a set of functional and operational performance requirements (Table 1). Each Functional and Operational Requirement (F&OR) requirement is then associated with a WPCS function that is subsequently allocated to one of the WPCS subsystems as presented in Fig. 3.

Table 1. Mapping of Project Functional and Operational Requirements (F&OR) allocated to WPCS subsystems. (NFD = no further decomposition of requirement within the F&OR document) (INEEL 2004a).

F&OR Requirement Number	Functional Requirement (verbatim F&OR)	Operational/ Performance Requirement (verbatim F&OR)	WPCS Function	WPCS Subsystem Allocation
1.1.2.3-1 (NFD)	The MGR shall identify the unsealed WP.	The MGR shall be capable of remotely identifying and recording the unique number of the WP (e.g., stamped).	*Admit the WP into WP Station*—The WP will be placed in the closure area and secured in one station suitable for subsequent operations.	Actual movement of the WPs into and out of the WPCS is a function controlled by others.
			Identify the WP—The WP and all components of the WP, including lids, caps, and spread rings, will be checked for proper identification before any operations are performed on them.	Data Management Subsystem
1.1.2.3-2a	The MGR shall perform WP sealing operations.	The process shall meet overall facility throughput. . . .	*Handle Closure Cell Materials*—Closure of the WPs requires handling operations: moving, storing, and transferring the closure lids and purge port cap into the closure cell and positioning them for welding; transferring tools into and out of the closure cell; properly indexing materials; and transferring equipment into gloveboxes for maintenance or repair.	Material Handling Subsystem
			Maintain Equipment Remotely—Because of the radiation environment within the closure cell, the WPCS is predominantly a remote operation, and techniques for servicing the equipment must be carefully planned into the design. Suitable equipment will be included in the WPCS for performing remote service. Decontamination capabilities are included under the functions of service, but the capability will be limited, based on the restriction of use of free liquids in the cell.	Miscellaneous Tool Subsystem

Table 1. (cont'd).

F&OR Requirement Number	Functional Requirement (verbatim F&OR)	Operational/ Performance Requirement (verbatim F&OR)	WPCS Function	WPCS Subsystem Allocation
1.1.2.3-2b	The MGR shall perform WP sealing operations	The WP seal shall not breach during normal operations or during credible preclosure event sequences.	Overall WPCS function	All WPCS Subsystems
1.1.2.3-2c	The MGR shall perform WP sealing operations	In conjunction with the other natural and engineered barriers, The WP seal shall limit the transport of radionuclides in a manner sufficient to meet long-term repository performance requirements.	Overall WPCS function	All WPCS Subsystems
1.1.2.3-2d	The MGR shall perform WP sealing operations.	The WP seal shall preclude moderator intrusion during preclosure and minimize the potential for moderator intrusion during the regulatory period (postclosure), which supports maintaining subcriticality.	Overall WPCS function	All WPCS Subsystems
1.1.2.3.2-1 (NFD)	The MGR shall weld lids and inerting caps.	The welding process shall be conducted in a safe, effective, and efficient manner to meet overall facility throughput . . . and weld requirements.	*Weld Lids and Purge Port Caps*—Each WP will be sealed by welding three lids in place. Welding will also be employed to seal the WP purge port cap once the container has been filled with an inert gas and plugged. Remote viewing and temperature sensing are required for control of the weld process. Capability for repair welding will be required, which necessitates grinding capability.	Welding Subsystem

Table 1. (cont'd).

F&OR Requirement Number	Functional Requirement (verbatim F&OR)	Operational/ Performance Requirement (verbatim F&OR)	WPCS Function	WPCS Subsystem Allocation
1.1.2.3.2-2 (NFD)	The MGR shall mitigate weld stresses.	Weld stresses shall be mitigated by imparting compressive residual stresses to an acceptable depth as defined in the Stress Corrosion Cracking AMR.	*Mitigate Weld Stresses*— A stress mitigation process is to be employed on the final closure weld to reduce residual weld-induced stresses.	Weld Stress Mitigation Subsystem (process type will be provided by BSC)
1.1.2.3.2-3 (NFD)	The MGR shall inert the sealed WP.	The amount of oxygen remaining in the WP shall be below predetermined limits.	*Inert the WP*—After the spread ring is seal welded, the interior of the WP is evacuated and backfilled with helium. Backfilling with inert gas (helium) replaces oxygen that could contribute to internal oxidation, enhance heat transfer after underground emplacement, and serve as the trace gas for leak testing the seal weld. The inerting function requires a method to determine the adequacy of the purge, a method to control the pressure of the backfill, and a method to sense that the inert gas is fully contained by the seal weld. If a leak is detected, it will be identified for repair.	Inerting Subsystem Leak Detection Subsystem
1.1.2.3-3 (NFD)	The MGR shall inspect WP seals.	The MGR shall be capable of inspecting WP seals remotely, using proven ASME-accepted Methods.	*Inspect Welds*—Nondestructive examination will be performed on the welds to ensure they are free of unacceptable defects. Methods included in the WPCS are quantitative visual, eddy current, and ultrasonic measurements.	Inspection Subsystem
1.1.2.3-4 (NFD)	The sealed WP shall provide conditions necessary to maintain the physical and chemical stability of the waste form.	The sealed WP environment shall provide conditions that restrict transport of radionuclides over the MGR period of performance.	Inert the WP.	Inerting Subsystem

Table 1. (cont'd).

F&OR Requirement Number	Functional Requirement (verbatim F&OR)	Operational/ Performance Requirement (verbatim F&OR)	WPCS Function	WPCS Subsystem Allocation
1.1.2.3-5 (NFD)	The MGR shall control WP closure systems operations.	The WP sealing process shall be remotely controlled in a manner that ensures safe, effective, and efficient WP closure.	*Control* WPCS *Operations*—The sequence of operations within the closure cell will be controlled to: 1) alleviate equipment conflicts within the cell; 2) ensure the reliability, accuracy, and consistency of the welds; and 3) protect operators and maintenance personnel from the environment inside the cell.	Control Subsystem

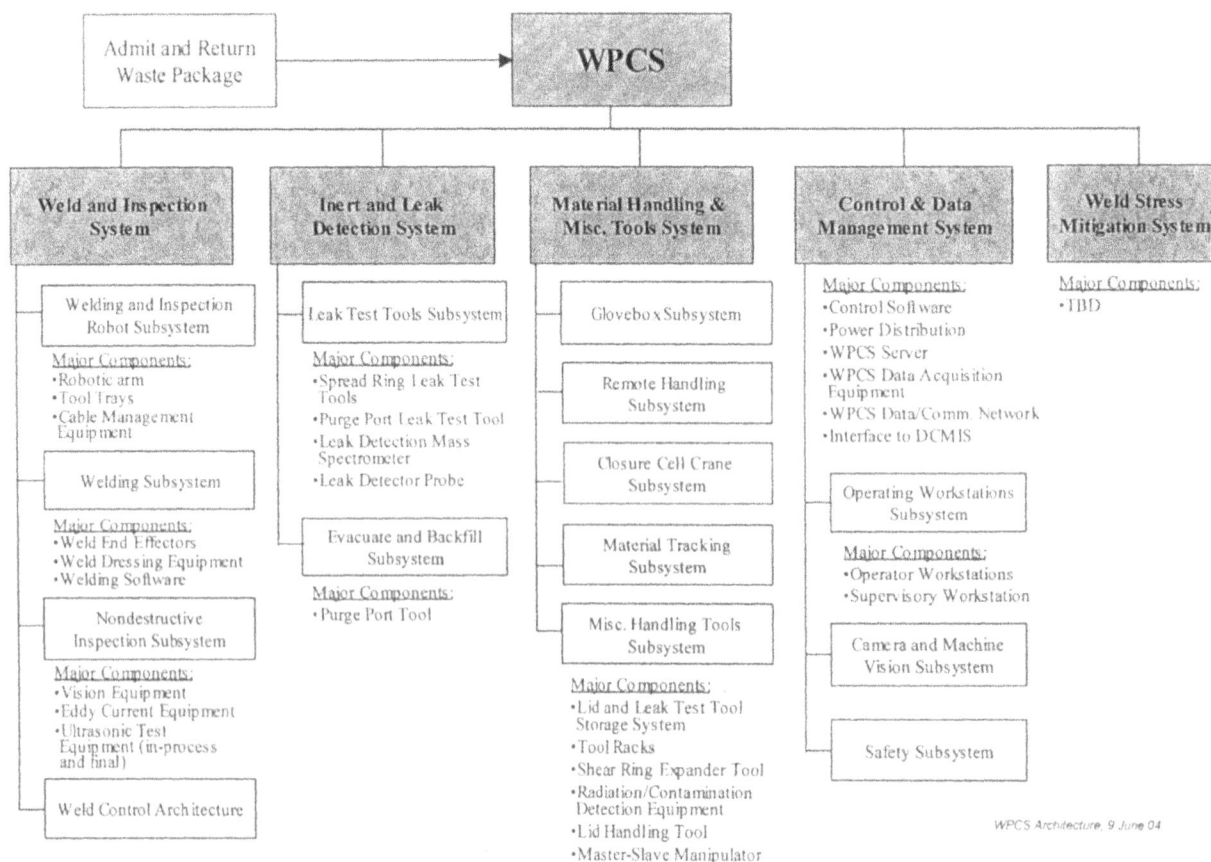

Fig. 3. The Waste Package Closure System architecture (INEEL 2004a).

Table 1 shows the complete mapping of upper tier F&OR requirements driving the major subsystems of the WPCS. The one-to-one function-to-subsystem mapping helps ensure compliance with requirements and simplifies interfaces as the design matures. Safety analyses that feed into an authorization basis for the WPCS will be performed by Bechtel SAIC Company (BSC). As they are completed, results will be included in the technical requirements documents.

Waste package closure system classification

Structures, systems, and components (SSCs) are classified in accordance with the definitions of the terms *Important to Safety* and *Important to Waste Isolation* provided by the USNRC (2003a).

The WPCS is not classified as Important to Safety nor Important to Waste Isolation as defined by Bechtel SAIC (2003).

Waste package closure system operational overview

The WPCS is located in the Yucca Mountain Project surface facility and comprises all the structures and equipment located in the closure cell; closure operating gallery; closure support area; closure maintenance cell; and maintenance area (Fig. 1). More closure cells may be located in a facility, but the layout and process will be identical in each one (i.e., additional cells are replicates). A simplified diagram of the sequence of events in the closure cell is shown in Fig. 4. Each of the major events depicted in Fig. 4 actually consists of many subsystem events.

The WPCS receives a waste package (WP) after it has been loaded with spent nuclear fuel/high-level waste (SNF/HLW). It will be positioned below a process opening in the operating level floor of the closure cell. Although there will be several different WP lengths, the tops of the various WPs will be at the same height with respect to the top of the floor in the closure cells. The WP will be unshielded, but the inner lid will be in place. The spread ring assembly will not be inserted, but will have been placed on the inner lid in the load cell before entering the closure area. The highest anticipated radiation field surrounding the WP will be about 1,500 rem/h (15 Sv/h) above the inner lid, and about 200 rem/h (2 Sv/h) to the side of the WP. The field will consist mainly of gamma radiation, but neutrons may also be emitted. Because of the high-radiation field surrounding the WP, personnel will not be able to enter the closure cell when the WP is present in the cell. Thus, all closure operations and most maintenance in the closure cell will be performed remotely. The temperature is estimated to be about 200°F (93°C) on the inner lid.

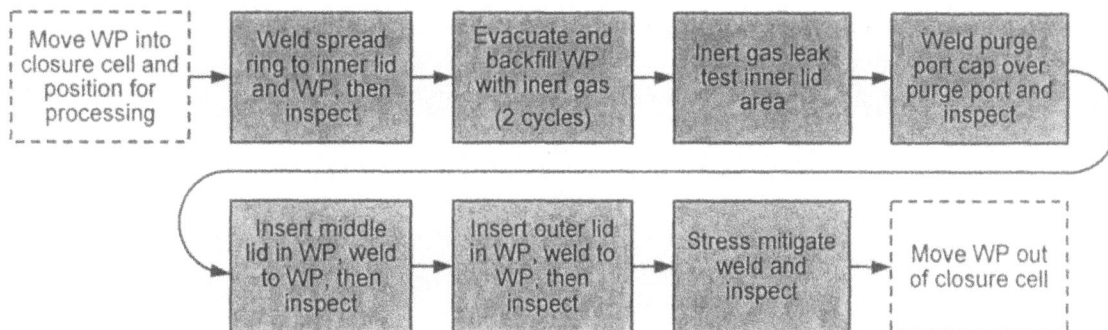

Fig. 4. Simplified sequence of events in the closure cell (INEEL 2004a).

The WP consists of two containers, one inside the other (Fig. 2): a Type 316 stainless steel inner vessel within an Alloy 22 shell. The specific identity of the WP and all components of the WP entering the closure area will be verified before entry. Visual examination by video will be performed before welding to ensure cleanliness and to verify position. A stainless steel inner lid will be inserted into the inner vessel before the WP enters the WP closure area (WPCA). Component temperatures will be measured by a thermocouple before tack-welding and before each weld pass. Weld interpass temperatures must not exceed 350°F (177°C) for the 316 stainless steel components, and 200°F (93°C) for the Alloy 22 components. The interpass temperatures were derived from vendor data. In the event these temperatures are exceeded, welding will not be performed, and suitable means will be taken to reduce the temperature of the components below the maximum temperature allowed before welding. All welding will be performed using the cold-wire gas tungsten arc-welding process.

A one-segment spread ring will be used to mechanically retain the inner lid. The spread ring will be tack-welded into position, and the tack welds will be visually examined and dressed, as necessary. A two-pass seal weld will be made between the spread ring and the inner lid; spread ring and inner vessel; and spread ring segment ends. This seal weld will be visually inspected. The inner vessel will then be evacuated and backfilled with helium through a purge port in the inner lid. The vacuum will be held for no less than 30 minutes. The inner vessel will then be evacuated and backfilled a second time with helium without the 30-minute vacuum hold time. The purity of the helium will also be checked before backfilling. A leak test with a mass spectrometer will be performed to ensure there is no helium in the region near the inner lid and associated seal welds. The purge port will then be plugged, leak tested, and covered by a purge port cap, which will be welded to the inner lid using a two-pass (minimum) seal weld. This seal weld will be visually inspected.

The middle Alloy 22 lid will be tack-welded to the Alloy 22 shell and dressed as necessary, followed by a multipass fillet weld. This weld will be visually and eddy-current inspected. A second Alloy 22 lid (the outer lid) will be placed in the Alloy 22 shell, tack-welded and dressed; and a multipass, full-thickness, narrow groove weld will be made joining the lid to the Alloy 22 shell. Ultrasonic inspection will be performed on the outer-lid weld following each weld pass. The cover pass will be inspected using visual, eddy current, and ultrasonic methods. Following the inspection, the weld will be stress-mitigated by inducing residual compressive stresses in the upper surface of the weld. A second inspection using visual, eddy current, and ultrasonic methods will be performed after the stress mitigation.

In the event any of the various inspections reveal an unacceptable indication, reasonable corrective actions will be taken in the closure cell. If repair of unacceptable indications in the closure cell is deemed not to be in the best interest of throughput, the WP will be sent to a designated location outside the closure cell for additional corrective actions.

The WPCS process has been designed to minimize the amount of time a WP stays in the closure cell in order to support the overall repository throughput goals. The closure process requires 44 hours to complete at 100% efficiency (no weld defects, operational delays, etc.). An estimated 60% efficiency (for maintenance, repairs, etc.) is assumed for the overall system, resulting in a throughput time of up to 70 hours for all closure operations.

The WPCA will be maintained in dry condition. Liquids are avoided in the WP closure cell to eliminate any chance of liquids entering a WP, and to reduce additional waste streams. Liquid inside a WP would jeopardize the corrosion resistance of the package and could also provide the conditions for a criticality event. Normally, no free liquids will be introduced into the closure cell—an exception being small controlled quantities of water used for ultrasonic inspection of the outer-lid weld and the stress mitigation process. Preclusion of liquids is especially important before completion of the seal welds on the inner lid.

WASTE PACKAGE CLOSURE SYSTEM REQUIREMENTS

The detailed requirements listed in Table 2 apply to the WPCS and may be fully or partially allocated to a subsystem or component for implementation. Many of the resulting requirements are qualitative in nature and have been allocated to other subsystems that will document quantitative measures.

Table 2. Technical Performance Requirements and Bases (INEEL 2004a).

WPCS Technical Requirement Number	WPCS Requirement	Performance Measure	Basis
TR-3.1-1	The WPCS will be designed to be compatible with the facility footprint.	Each WPCA will be a fixed area within the facility; additional cells will be duplicates of this area.	The WPCS must be within the same facility where loading occurs to prevent the risk of contamination spread outside an enclosure.
TR-3.1-2	Process cycle time without weld repairs is 44 hours.	Online operating efficiency shall be 60%, and welding operations shall be completed within 70 hours (44 h for optimum, 100%, efficiency) of the WP being secured in the closure cell.	BSC requirement.
TR-3.1-3	Each WPCS shall be capable of continuous operations.	Operations will be 24 h/day, 7 days/week, 365 days/year, except for scheduled maintenance and other scheduled and unscheduled events to be determined.	BSC requirement.
TR-3.1-4	The WPCS shall be capable of operations and maintenance for the facility life expectancy.	Operations will continue for 50-year postconstruction.	Safety Analysis Report.
TR-3.1-5	The WPCS shall record and store all pertinent operations data in an electronic database.	At minimum, records shall be compliant with NRC regulation.	The Project Team expects the NRC to require operational data to ensure adequate closure.
TR-3.1-6	The top center of the WP outer vessel shall be positioned in a predetermined location.	The top center of the WP must be: a) Between 7½ in (19 cm) and 12 in (30 cm) below the top surface of the second floor in the z direction. b) ±2 in (±5 cm) from the center of the process opening on the operating floor in the x and y directions. c) Within maximum tilt from horizontal ±½ in (±13 mm). See Figure 5.	BSC requirement.
TR-3.1-7	WP vibration shall be limited to TBD.		Facility design.

Boundaries and interfaces requirements and bases

The WPCS includes all equipment and operations within the WPCA that directly support closure of a waste package. Ancillary SSCs, such as Facility walls; floors; ceilings; doors; windows; lighting; HVAC (heating, ventilation, and air conditioning); electrical power; compressed gas supplies; radiological alarms; and equipment anchors and supports are specified and provided by the Facility design groups. The Facility design groups are responsible for moving the WP into the WPCA; securing it within the required tolerances; and moving the WP out of the WPCA once closure is completed. The Facility is also responsible for the moving door between the closure cell and the maintenance area. The following sections include a discussion of the WPCS requirements that must be met by the Facility.

Physical location and layouts

The location of the WP in the WP station will be controlled by the Facility. It will be positioned within a 9-ft (2.7-m)-diameter process opening: the center point 24 ft (7.2 m) from one side wall; 16 ft (4.8 m) from the opposite side wall; and 10 ft (3 m) from the operating gallery wall. The top of the WP must be within 7 ½ and 12 in (19 and 30 cm) below the top surface of the second floor; ±2 in (5 cm) in the X and Y direction from the center of the process opening, and less than ±½ in (1.25 cm) in flatness (Fig. 5). The WPCS is located between the SNF/HLW transfer cell and the WP loadout cell. Several areas are used to complete the WPCS operations: the WP station on the ground floor; the closure cell; the WP closure maintenance cell; the closure support area; the closure operating gallery on the operating floor; and the maintenance area on the upper floor. The current design is shown in Figs. 1, 6, and 7.

All operations on the WP (welding, NDE, inerting, stress mitigation) will occur in the closure cell. No access to the closure cell will be allowed when a WP is present, but a plate will be available to cover the process opening if needed during non-operating periods. Frequent maintenance activities will be performed in the glovebox, located in the closure support area. Weld tool trays will be moved from the closure cell into the glovebox by way of the transfer tunnel for change-out of weld wire and calibration.

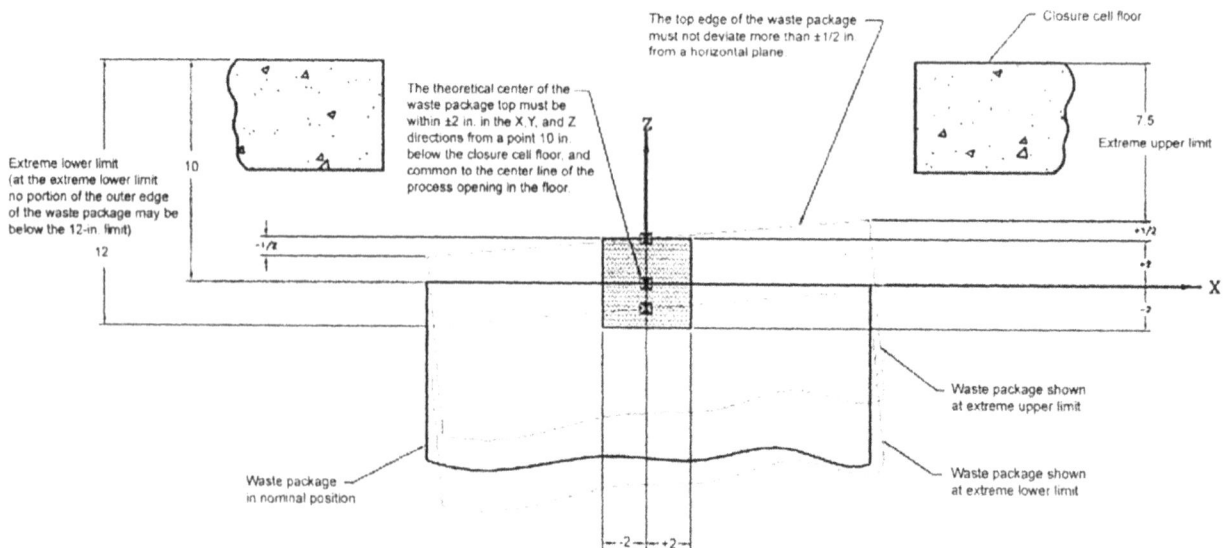

Fig. 5. Waste package position relative to the process opening in the closure cell floor (INEEL 2004a).

46

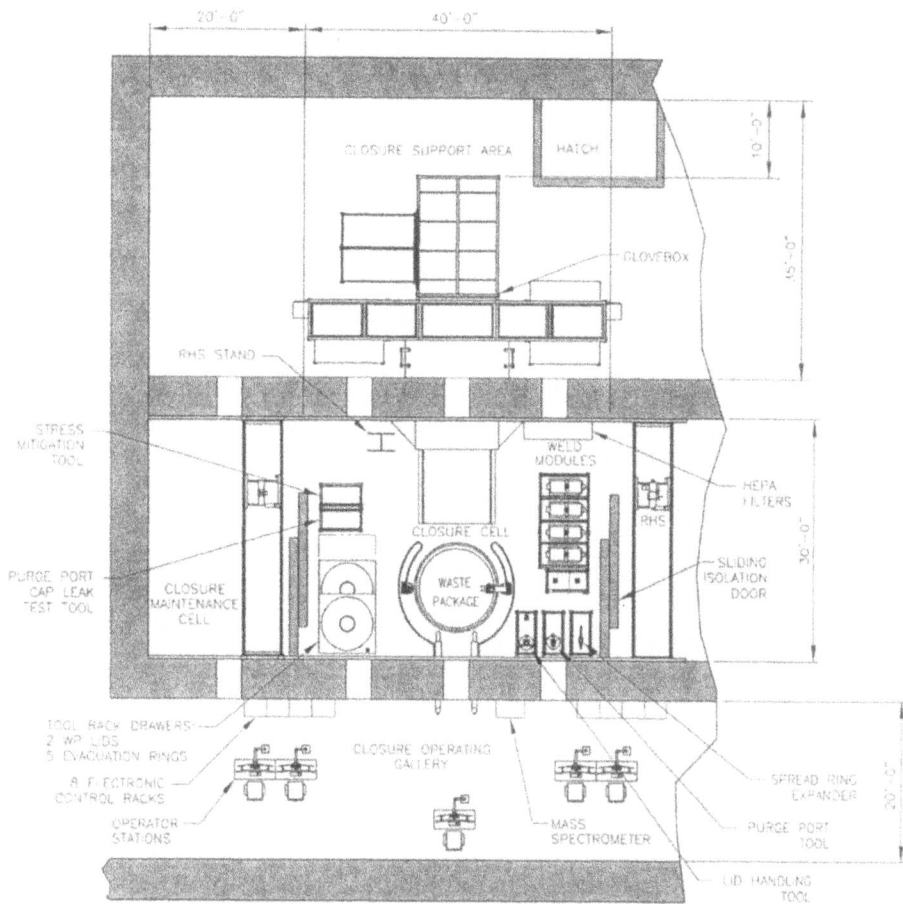

Fig. 6. Plan view of the operating floor of the Waste Package Closure System (INEEL 2004a).

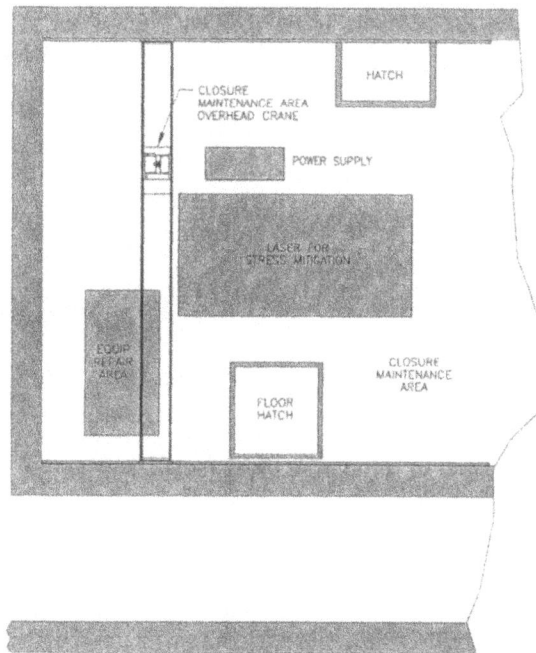

Fig. 7. Plan view of the upper floor of the Waste Package Closure System (INEEL 2004a).

The purge port tool and stress mitigation equipment may also be serviced in the glovebox. Transfer of materials (lids, etc.) into the closure cell will also be achieved using the transfer tunnel. Larger equipment (leak detection tool, master slave manipulators, cranes) can be routinely serviced in the WP closure maintenance area, at estimated intervals of 6 to 12 months. Large process support equipment, such as laser peening (if selected) for the stress mitigation process and vacuum equipment, will be located in the maintenance area on the upper floor. Hatches to the closure cell and closure support area allow the maintenance area crane to move equipment into the maintenance area for repair if necessary.

System reliability features

The system reliability features are listed in Table 3. Commercial equipment and products should be used wherever practical to reduce cost and schedule, and to improve maintainability of the overall WPCS. Commercial products have the advantage to be readily available, and have shorter delivery times; known reliability; available spare parts; and generally-lower cost.

Table 3. System Reliability Features (INEEL 2004a).

WPCS Technical Requirement Number	Requirement	Allocated to
TR-3.4-1	Define reliability and availability requirements for each WPCS subsystem.	WPCS subsystems
TR-3.4-2	Identify preferred failure modes for each WPCS subsystem.	WPCS subsystems

System control features

Individual components (such as welding or inerting) will have individual control features. An overarching supervisory control system (SCS) will also be employed to ensure: 1) that each individual operation is coordinated with the other operations and 2) that conflicts are avoided. Some interface with the digital control and management information system (DCMIS) will be required.

System operation requirements

The system operation requirements cover the design disciplines and include additional requirements which are necessary for the proper, safe, and efficient operation of the WPCS. Tables 4-18 list the system operation requirements for the following:

1. Facility/Building
2. Communication
3. Civil and Structural
4. Mechanical and Materials
5. Chemical and Process
6. Electrical Power
7. Instrumentation and Control
8. Computer Hardware and Software

9. Operating Environment and Natural Phenomena
10. Fire Protection
11. Radiation Hazards
12. Industrial Hazards requirements
13. ALARA
14. Human Interface
15. Environmental Management

Table 4. Facility/Building Requirements (INEEL 2004a).

WPCS Technical Requirement number	Requirement	Allocated to
TR-3.6.1-1	Rooms will be provided to perform closure operations, storage out of cell, maintenance, and control operations.	Facility Design
TR-3.6.1-2	The facility structure shall provide shielding for personnel and, in some cases, equipment.	Facility Design
TR-3.6.1-3	There shall be openings for passing tools, parts, and equipment between the closure cells and the closure support area. Must be compatible with requirement TR-3.6.3-1.	Facility Design
TR-3.6.1-4	Shielded windows shall be provided for viewing operations and maintenance.	Facility Design
TR-3.6.1-5	Facilities shall be designed to provide protection against optical hazards.	Facility Design
TR-3.6.1-6	There shall be provisions for anchoring tools and equipment to the facility structure.	Facility Design
TR-3.6.1-7	The facility structure shall be capable of supporting the loads imposed by the tools and equipment anchored to the facility structure.	Facility Design
TR-3.6.1-8	The closure support area, closure operating gallery, and the maintenance area shall be designed to allow continuous occupancy.	Facility Design
TR-3.6.1-9	The facility walls and ceiling shall provide shielding from the design basis source term within the closure cell to all adjacent areas (including areas not normally occupied).	Facility Design
TR-3.6.1-10	The closure cell shield doors shall provide total radiological shielding isolation between the maintenance areas and the adjoining closure cell to allow personnel entry while a WP is present in the closure cell.	Facility Design
TR-3.6.1-11	The closure cell (walls, doors, etc.) shall be capable of being readily decontaminated.	Facility Design
TR-3.6.1-12	The closure cell shield door controls shall have an Ethernet/TCP/IP interface to the WPCS control system.	Facility Design
TR-3.6.1-13	The WP shall be placed in the WPCA to be sealed.	Facility Design

Table 4. (cont'd).

WPCS Technical Requirement number	Requirement	Allocated to
TR-3.6.1-14	Personnel access shield doors shall be provided into the closure maintenance area from the closure support area Shielding for personnel in the closure support area must be maintained during operations.	Facility Design
TR-3.6.1-15	Personnel access shield doors shall be designed for manual operation.	Facility Design
TR-3.6.1-16	The personnel access shield doors shall form an airlock with an outer contamination control door that is nonshielding.	Facility Design
TR-3.6.1-17	The closure maintenance areas shall be equipped with service utilities to perform all intended maintenance functions (breathing air, plant air, decontamination services, electrical power).	Facility Design
TR-3.6.1-18	The closure cell support area elevator lift shall provide access to all three levels of WPCA.	Facility Design
TR-3.6.1-19	The closure cell support area elevator lift shall be able to accommodate TBD size and weights.	Facility Design

Table 5. Communication Requirements (INEEL 2004a).

WPCS Technical Requirement number	Requirement	Allocated to
TR-3.6.2-1	Video interfaces shall be provided for integration into a control room environment.	Digital Control and Management Information System WPCS Control and Data Management Subsystem
TR-3.6.2-2	An Ethernet/TCP/IP interface shall be provided at the closure cell control system for central control system communications.	Digital Control and Management Information System
TR-3.6.2-3	Communications between the various subsystems and the WP Closure Control System shall employ the Open Systems Interconnection Reference Model.	Digital Control and Management Information System
TR-3.6.2-4	In-cell electrical connectors must meet the following criteria: Constructed of materials compatible with the radiation environment. Bulkhead connectors shall be capable of being connected or disconnected by a remote manipulator (except where otherwise stated).	Digital Control and Management Information System WPCS Control and Data Management Subsystem

Table 5. (cont'd).

WPCS Technical Requirement number	Requirement	Allocated to
TR-3.6.2-4	Unused bulkhead connectors will be capped. Instrumentation wiring will be made of a gauge and with shielding appropriate to the application. Power lines will run in conduits separate from instrumentation and control lines.	

Table 6. Civil and Structural Requirements (INEEL 2004a).

WPCS Technical Requirement Number	Requirement	Allocated to
TR-3.6.3-1	The WPCS shall provide for entry and removal of materials and equipment, while preventing releases of radioactive and hazardous contaminants above the threshold limits to the environment. Must be compatible with Requirement TR-3.6.1-3.	WPCS glovebox component of the Material Handling Subsystem
TR-3.6.3-2	The waste package closure maintenance areas (operating floor) shall be equipped with manipulators for remotely servicing equipment.	WPCS master-slave manipulator component of the Material Handling Subsystem
TR-3.6.3-3	There shall be provisions for parking the overhead handling devices out of the radiation streaming from the top of the waste package.	WPCS Material Handling Subsystem

Table 7. Mechanical and Materials Requirements (INEEL 2004a).

WPCS Technical Requirement Number	Requirement	Allocated to
TR-3.6.4-1	A vacuum system shall be provided for evacuation of the WP.	Plant Services WPCS Evacuate and Backfill Subsystem
TR-3.6.4-2	Use of materials that degrade in high radiation fields, such as polymers and many lubricants, shall not be used unless a suitable substitute is not available.	All applicable WPCS subsystems and components

51

Table 7. (cont'd).

WPCS Technical Requirement Number	Requirement	Allocated to
TR-3.6.4-3	The HVAC systems shall have the capacity to maintain a constant air temperature in the occupied areas.	Industrial HVAC System
TR-3.6.4-4	All equipment within the closure cells and WP closure maintenance areas shall be serviceable remotely, in the glovebox, in the WP closure maintenance area, or capable of being decontaminated for removal and hands-on service.	All WPCS subsystems and components
TR-3.6.4-5	The waste package closure area (WPCA) shall be maintained in a dry condition. Normally, no free liquids shall be introduced into the WPCA. Equipment containing liquids shall be designed in a manner to reasonably preclude leakage of those fluids into the cell. This is especially important for equipment used before completion of the seal welds on the inner lid. Small amounts of free controlled liquids may be released into the WPCA directly associated with ultrasonic inspection and stress mitigation of the narrow-groove closure weld between the outer lid and the alloy 22 shell.	All applicable WPCS subsystems and components

Table 8. Chemical and Process Requirements (INEEL 2004a).

WPCS Technical Requirements Number	Requirement	Allocated to
TR-3.6.5-1	Compressed gases (helium, air, and argon) shall be available in sufficient supply to perform closure operations. Quantities will be defined in the CDDs.	Plant Services All WPCS subsystems
TR-3.6.5-2	Vacuum interface shall be available for the grinding/cleaning system.	Plant Services WPCS Welding Subsystem
TR-3.6.5-3	Service utilities will be required in the WPCA. Utility type and quality will be defined in the CDDs.	Plant Services All plant subsystems

Table 9. Electrical Power Requirements (INEEL 2004a).

WPCS Technical Requirement Number	Requirement	Allocated to
TR-3.6.6-1	Each closure cell shall have 480-V, 3-phase, [TBD]-kVA electrical power.	Electrical System
TR-3.6.6-2	240-V, 2-phase, [TBD]-kVA power panels will be available to each operating gallery and support area, as defined by a single-line electrical drawing.	Electrical System
TR-3.6.6-3	Total demand is expected to be [TBD]-kVA for all electrical equipment, excluding the welding equipment.	Electrical System
TR-3.6.6-4	The power requirements for the facility to support the welding equipment are [TBD] kVA at [TBD] V alternate current.	Electrical System
TR-3.6.6-5	Uninterruptible facility power (480 V, 3-phase, [TBD]-kVA) will be available to each weld system, as defined by a single-line electrical drawing.	Electrical System

Table 10. Instrumentation and Control Requirements (INEEL 2004a).

WPCS Technical Requirement Number	Requirement	Allocated to
TR-3.6.7-1	The WPCS shall provide its own individual control system.	WPCS Control Subsystem
TR-3.6.7-2	Camera outputs shall provide a National Television Standards Committee output signal. Any cameras that provide other signal formats will be addressed in the applicable CDD.	WPCS Control Subsystem
TR-3.6.7-3	The WPCS control electronics shall provide emergency shutdown, safety locks, alarms or warnings, fail safe planning, and other functions as needed to provide a safe system for the operator, support personnel, the public, equipment, and environment.	WPCS Control Subsystem
TR-3.6.7-4	Electronic components shall be located outside of the hot cell whenever possible.	WPCS Control Subsystem

WPCS Data Management Subsystem

WPCS Inspection Subsystem |

Table 11. Computer Hardware and Software Requirements (INEEL 2004a).

WPCS Technical Requirement Number	Requirement	Allocated to
TR-3.6.8-1	Any computer programs used within the control processes for the WPCS shall be controlled in accordance with a software Quality Assurance plan.	All applicable WPCS subsystems

Table 12. Operating Environment and Natural Phenomena Requirements (INEEL 2004a).

WPCS Technical Requirement Number	Requirement	Allocated to
TR-3.6.9-1	The WPCS critical lift equipment shall be designed to operate during and following design-basis seismic events according to ICC (2000).	WPCS Material Handling Subsystem
TR-3.6.9-2	The WPCS shall be designed to withstand the effects of Yucca Mountain natural phenomena in accordance with ICC (2000).	All WPCS subsystems and components

Table 13. Fire Protection Requirements (INEEL 2004a).

WPCS Technical Requirement Number	Requirement	Allocated to
TR-3.6.10-1	The WPCS shall be designed, constructed, operated, and maintained in a manner that minimizes the potential for fires and explosions.	All WPCS subsystems and components

Table 14. Radiation Hazards Requirements (INEEL 2004a).

WPCS Technical Requirement Number	Requirement	Allocated to
TR-3.6.11-1	Design of the WPCS shall accommodate the radiation fields shown in Figures 8-10.	All in-cell WPCS subsystems and components.
TR-3.6.11-2	Design of the WPCS shall accommodate decontamination of the cell equipment.	All in-cell WPCS subsystems and components.

Table 15. Industrial Hazards Requirements (INEEL 2004a).

WPCS Technical Requirement Number	Requirement	Allocated To
TR-3.6.12-1	The WPCS design shall provide for protection of personnel from electrical hazards.	All WPCS subsystems and components
TR-3.6.12-2	The WPCS shall be designed to reduce the risk of exposure of personnel to hazardous gases.	Welding subsystem Purge and Backfill Subsystem
TR-3.6.12-3	Hazards associated with lifts of heavy loads shall be addressed by adherence to DOE (2004) or equivalent.	Welding Subsystem Material Handling Subsystem
TR-3.6.12-4	Emergency stop controls and lockout/tagout shall be provided for welding equipment; welding power supplies; robotic, crane, and hoisting equipment; shield doors; etc.	Welding Subsystem Material Handling Subsystem
TR-3.6.12-5	The WPCS shall ensure protection of workers in accordance with OSHA (2003) or equivalent.	All WPCS subsystems and components
TR-3.6.12-6	The asphyxiation hazard associated with the shielding gas used in the weld process shall be considered in the HVAC system design.	Industrial HVAC system

Table 16. As Low as Reasonably Achievable Requirements (INEEL 2004a).

WPCS Technical Requirement Number	Requirement	Allocated to
TR-3.6.13-1	The WPCS design shall implement remote operations and maintenance to the extent feasible.	All in-cell WPCS subsystems and components
TR-3.6.13-2	The WPCS shall apply as low as reasonably achievable (ALARA) principles of exposures to materials (radioactive or hazardous) to ensure worker safety.	All WPCS subsystems and components
TR-3.6.13-3	The WPCS equipment shall be designed and fabricated to facilitate decontamination before maintenance or to reduce radiation fields for in-cell maintenance activities.	All WPCS subsystems and components

Table 17. Human Interface Requirements (INEEL 2004a).

WPCS Technical Requirement Number	Requirement	Allocated to
TR-3.6.14-1	Manned workstations shall be ergonomically designed.	WPCS Control Subsystem
TR-3.6.14-2	Hazards associated with repetitive motions at workstation terminals will be addressed in the design of the workstations and the application of administrative controls.	WPCS Control Subsystem

Table 18. Environmental Management Requirements (INEEL 2004a).

WPCS Technical Requirement Number	Requirement	Allocated to
TR-3.6.15-1	All WPCS elements shall be operated in a safe and environmentally sound manner.	All WPCS subsystems and components
TR-3.6.15-2	The WPCS shall produce no mixed hazardous waste.	All WPCS subsystems and components
TR-3.6.15-3	The WPCS shall minimize the generation of all waste materials.	All WPCS subsystems and components

Heat generated by the contents of the WP and by the welding operations must be considered in the operating environment analysis. Ionizing radiation from the WP will be significant, and is a major concern in the equipment design—particularly for the electronic items. Specific radiation fields will be estimated based on the fuel inventory and the cell design. Other cell environment issues, such as electromagnetic noise, will be addressed in the individual component design descriptions (CDDs) as the design develops. Natural phenomena risks are discussed in *Project Design Criteria Document* (Minwalla 2003). Since the WPCS is not important to safety and not important to waste isolation, the seismic criteria are according to *International Building Code 2000* (ICC 2000). The WP enters the WPCA loaded with SNF or HLW material. Although the inner lid will have been placed into position, there will be considerable radiation in the form of gamma radiation and some neutron radiation that will be emitted from the top of the waste package. This radiation is of sufficient intensity to pose a significant operational hazard to various components of the closure equipment including electronics, insulation, lubricants, and other materials. Both their lifetimes and their ability to operate (in the case of video cameras, for example) are issues that need to be addressed. As a general rule, although radiation-hardened components can be used in some cases, they are expensive. A solution is to either shield or locate such components out of the radiation areas. The highest radiation field is located directly over the top of the WP. Consequently, as many components as possible will be placed to the side of the WP in a location that can be easily shielded.

Current designs for the closure cell do not include fire suppression. If a fire should occur within the cell, it will burn itself out. Other areas in the WPCA will have fire suppression.

Figures 8-10 show the radiation dose rates expected for different segments around the WP, based on the 21-PWR (pressurized water reactor) fuel (CRWMS M&O 2000). The rates given are the average for the zone depicted. Figure 8 shows the bounding values, which represent the expected maximum for the 21-PWR fuel. Figure 9 illustrates the average values for the same fuel. Both the bounding and average values are evaluated when all three lids are installed on the WP. Figure 10 shows the expected dose rates at the surface of the WP top section as each lid is added to the WP. Work is currently in progress to simulate the radiation fields with a cell floor and ceiling in place to obtain realistic dose rates for equipment installed in the cell.

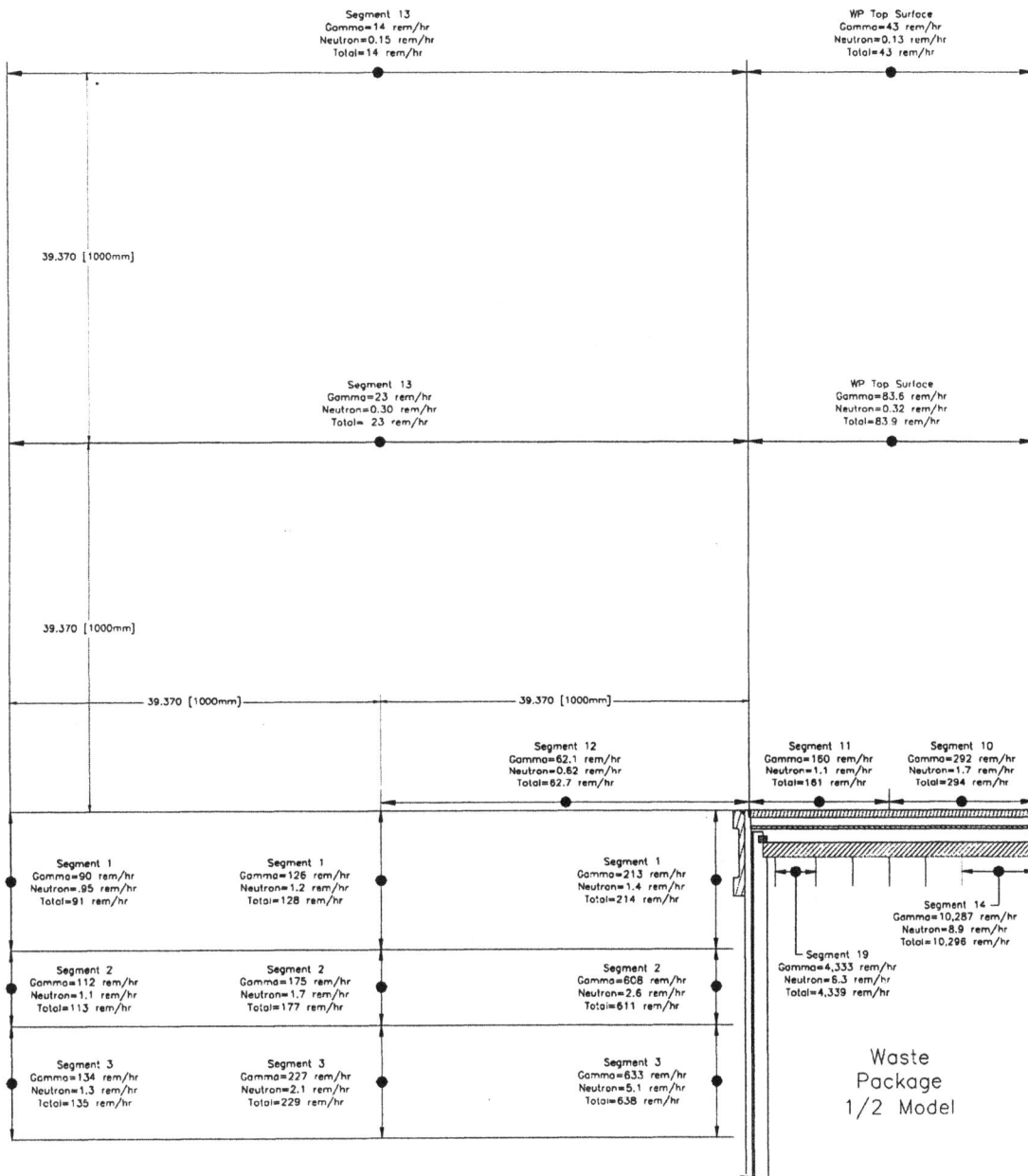

Fig. 8. Maximum dose rates (INEEL 2004a).

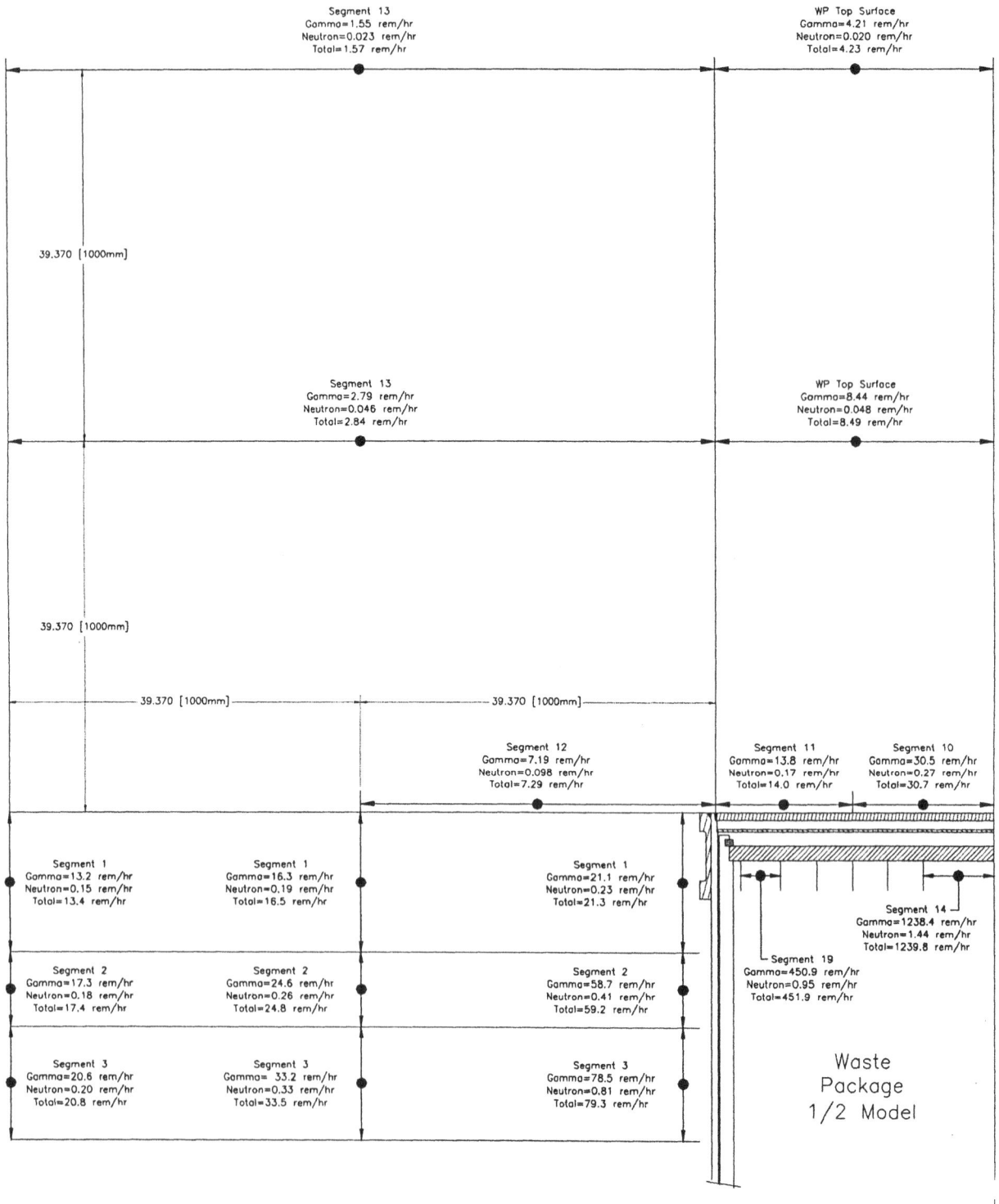

Average Dose Rates
Data Taken from: Dose Rate Calculation for the 21-PWR UCF
Waste Package (Tables 37, 38, 39 & 40)
Document Identifier: CAL-UDC-NU-000002

Segment 13
Gamma=1.55 rem/hr
Neutron=0.023 rem/hr
Total=1.57 rem/hr

WP Top Surface
Gamma=4.21 rem/hr
Neutron=0.020 rem/hr
Total=4.23 rem/hr

39.370 [1000mm]

Segment 13
Gamma=2.79 rem/hr
Neutron=0.046 rem/hr
Total=2.84 rem/hr

WP Top Surface
Gamma=8.44 rem/hr
Neutron=0.048 rem/hr
Total=8.49 rem/hr

39.370 [1000mm]

39.370 [1000mm]

39.370 [1000mm]

Segment 12
Gamma=7.19 rem/hr
Neutron=0.098 rem/hr
Total=7.29 rem/hr

Segment 11
Gamma=13.8 rem/hr
Neutron=0.17 rem/hr
Total=14.0 rem/hr

Segment 10
Gamma=30.5 rem/hr
Neutron=0.27 rem/hr
Total=30.7 rem/hr

Segment 1
Gamma=13.2 rem/hr
Neutron=0.15 rem/hr
Total=13.4 rem/hr

Segment 1
Gamma=16.3 rem/hr
Neutron=0.19 rem/hr
Total=16.5 rem/hr

Segment 1
Gamma=21.1 rem/hr
Neutron=0.23 rem/hr
Total=21.3 rem/hr

Segment 14
Gamma=1238.4 rem/hr
Neutron=1.44 rem/hr
Total=1239.8 rem/hr

Segment 2
Gamma=17.3 rem/hr
Neutron=0.18 rem/hr
Total=17.4 rem/hr

Segment 2
Gamma=24.6 rem/hr
Neutron=0.26 rem/hr
Total=24.8 rem/hr

Segment 2
Gamma=58.7 rem/hr
Neutron=0.41 rem/hr
Total=59.2 rem/hr

Segment 19
Gamma=450.9 rem/hr
Neutron=0.95 rem/hr
Total=451.9 rem/hr

Segment 3
Gamma=20.6 rem/hr
Neutron=0.20 rem/hr
Total=20.8 rem/hr

Segment 3
Gamma=33.2 rem/hr
Neutron=0.33 rem/hr
Total=33.5 rem/hr

Segment 3
Gamma=78.5 rem/hr
Neutron=0.81 rem/hr
Total=79.3 rem/hr

Waste
Package
1/2 Model

Fig. 9. Average dose rates (INEEL 2004a).

58

Average
Dose Rates

Bounding
Dose Rates

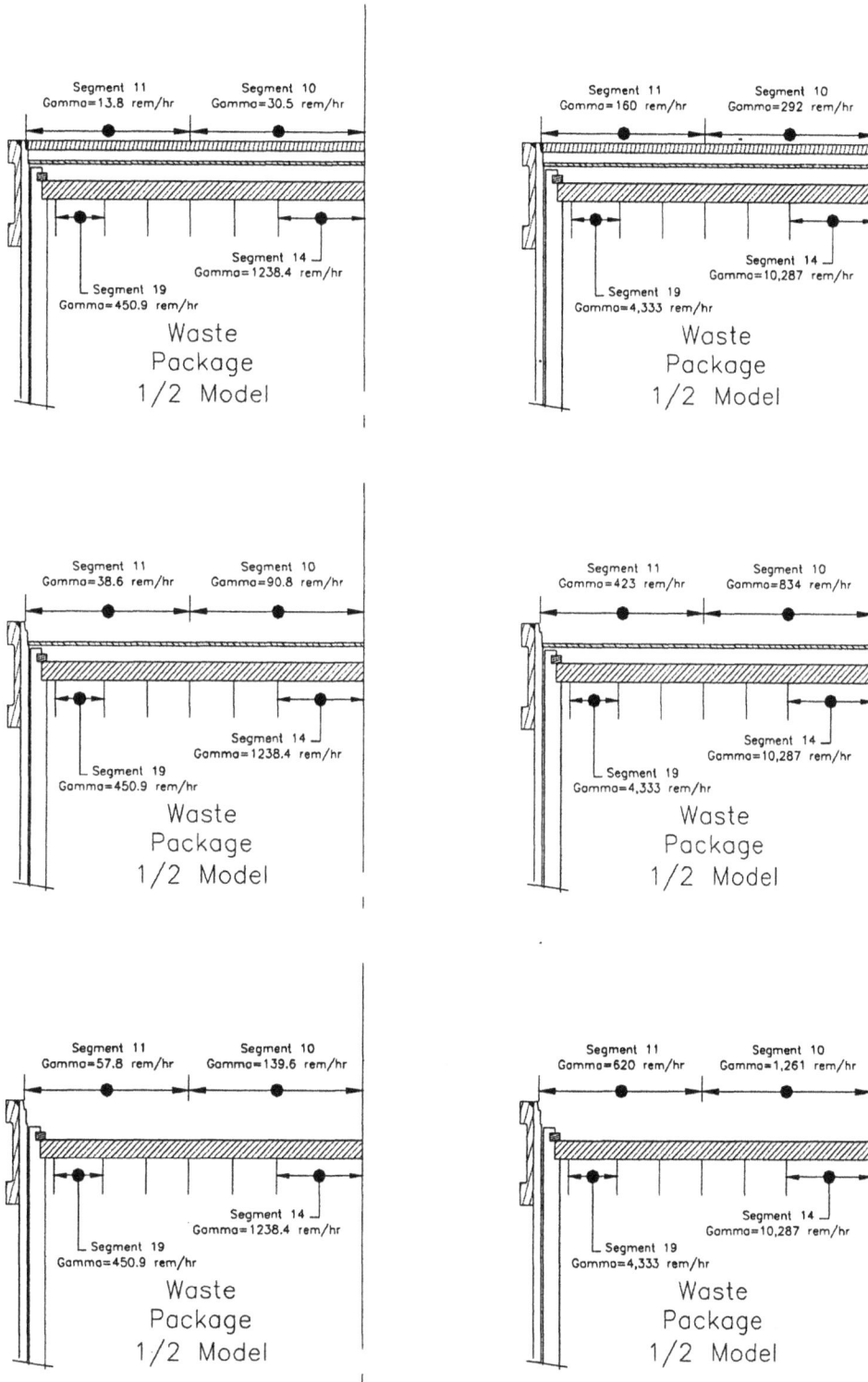

Fig. 10. Separate lid dose rates (INEEL 2004a).

59

No free liquids will be deliberately released in the closure cells except for ultrasonic inspection of the outer lid closure weld, and possibly the stress mitigation process. The containment of liquids for this process will be designed into the system.

The USNRC (2003b) requires that the safety analysis report describe design considerations that are intended to facilitate permanent closure and decontamination or decontamination and dismantlement of surface facilities.

Testing and maintenance requirements

The Testability (Table 19) and Maintanance (Table 20) requirements presented in this section relate to the design of the system, as opposed to operational testing and maintenance requirements.

Design Life Category 1 refers to equipment that can be easily replaced and not cause downtime to the cell. This equipment must be easily removable by remote means for transport to the glovebox, where it can be serviced or replaced. Equipment in this category must function for at least 100 hours before failure to ensure that a WP can be completed without any equipment change-out. Equipment will receive preventative maintenance or be replaced, as required, within the glovebox. An example of equipment in this category is the o-rings on the welding torch, which will deteriorate in the high-radiation field and will require periodic replacement in the glovebox.

Table 19. Testability Requirements (INEEL 2004a).

WPCS Technical Requirement Number	Requirement	Allocated to
TR-3.7.1-1	Before the final selection of subsystems that compose the WPCS, INEEL shall demonstrate the subsystems.	All WPCS subsystems
TR-3.7.1-2	System testing shall be completed after construction of the WPCS and before the demonstration.	All WPCS subsystems

Table 20. Maintenance Requirements (INEEL 2004a).

WPCS Technical Requirement Number	Requirement	Allocated to
TR-3.7.2-1	Equipment specifications shall require the equipment or component vendor (or designer for unique items) to identify preventive maintenance and anticipated repair instructions.	All WPCS subsystems and components
TR-3.7.2-2	Equipment designs must include mean time between failures and mean time to repair for all standard WPCS equipment.	All WPCS subsystems and components

Design Life Category 2 refers to equipment that cannot be serviced within the glovebox. This equipment will be moved to the closure maintenance area for service, which will require manned entry into the area to conduct the work. It is not expected that entry into the closure maintenance cell will require shutdown of closure cell operations. Equipment in Category 2 must function for a minimum of 6 months before failure. Maintenance will be performed twice a year, and all equipment in this category will receive preventative maintenance or be replaced, as required. An example of equipment in this category is the large spread ring leak test tool, which may require seal replacement at 6-month intervals.

Design Life Category 3 refers to equipment that is fixed in place within the weld cell and cannot be moved for maintenance, or would require shutdown of closure cell operations even if it could be moved. Servicing of equipment in this category requires shutdown of cell operations and manned entry into the closure cell area. Equipment in this category must function for at least 1 year before failure. Maintenance within the closure cell will be performed yearly, and all equipment in this category will receive preventative maintenance or will be replaced, as required. An example of equipment in this category is the cart track, which is too large to be easily removed.

DESIGN GUIDELINES

Table 21 presents guidelines for the design of the WPCS for the Yucca Mountain Project. These guidelines were developed during the conceptual and preliminary design stages of the WPCS. The difference between these guidelines and the system requirements documented above is: though the guidelines are presented in imperative/requirement language, the designers may determine the extent to which these guidelines are applied, based on cost; schedule; risk; and other performance parameters. Many of these guidelines will be translated into subsystem or component requirements, based on the outcome of the WPCS prototyping effort. Until that occurs, the designers are encouraged to find the optimal application of these guidelines through experimental activities.

Table 21. Design Guidelines (INEEL 2004a).

No.	Design Guideline	Applies to
1	Radiation shielding windows through the closure cell shield wall shall be provided at strategic locations.	Facility
2	Shielding windows shall conform to the ASTM Guide for radiation-shielding window components used in hot cells.	Facility
3	The remote maintenance areas shall be provided with adequate lighting.	Facility
4	The remote maintenance areas shall be equipped with the necessary working platforms to perform all intended maintenance and repair activities.	Facility
5	The facility shall be designed for safe operation in accordance with ANSI (1973 or latest edition) where applicable.	Facility
6	The facility shall be designed for safe operation in accordance with laser safety requirements.	Facility
7	HVAC systems shall provide adequate air exchanges to prevent buildup of hazardous gases.	Facility
8	WPCS control systems shall be designed for ease of operability, remote maintenance, and decontamination.	WPCS

Table 21. (cont'd).

No.	Design Guideline	Applies to
9	The WPCS shall maintain equipment remotely to reduce risk of contamination spread and personnel exposure to radiation.	WPCS
10	The lighting shall be remotely replaceable. Lighting is necessary for vision of operations and must be replaceable at any time without personnel entry.	Facility
11	Materials, tools, and equipment in the closure cell shall not pose a risk of contaminating the WP with foreign elements that could promote corrosion or poor welds (e.g., sulfur compounds, hydrocarbons, zinc, and halides.)	All WPCS subsystems
12	References for selecting materials for radiation environments, for example Vandergriff (1990) and van de Voorde and Restat (1972), shall be considered in designing WPCS tooling and equipment.	All WPCS subsystems and components exposed to radiation environment
13	From a decommissioning aspect, equipment designed to fit in a 55-gal drum (whole or after dismantling) is preferred.	All WPCS subsystems and components
14	The closure cell equipment shall be capable of being readily decontaminated and shall include features permitting in-cell decontamination of in-cell items.	WPCS
15	Where practical, tools and equipment shall be designed to accommodate the different sizes of waste packages and perform multiple tasks to minimize the number of tools. Fewer tools to perform the same task simplify the operation and allow fewer movements in the cell, resulting in decreased risk of failure.	WPCS
16	When designing the WPCS tooling and equipment, consider the guidelines in DOE (1999) and AGS (1998).	WPCS
17	Commercial equipment and products shall be used wherever possible to reduce cost and schedule and to improve reliability and maintainability of the overall YMP system. Commercial products are readily available and have shorter delivery times, known reliabilities, and spare parts availability, and are generally lower in cost.	WPCS
18	The preventive maintenance and anticipated repair frequency shall be considered in the final selection of equipment vendors or designs. Consideration shall include overall project decisions regarding equipment availability, overall system availability, compatibility with the WPCA, and identification of spares.	WPCS
19	Stress analysis of items where required shall conform to an applicable nationally recognized design specification (i.e., Aluminum Association's design manual, the American Institute of Steel Construction (AISC) Manual of Steel Construction)	WPCS
20	Lifting devices shall meet the requirements of the DOE (2004).	WPCS

WELDING AND INSPECTION SYSTEM

THE WASTE PACKAGE AND THE WELDING AND INSPECTION SYSTEM

The following material is derived from the TFR-283, Component Design Description: Welding and Inspection System (INEEL 2004b) document.

During repository operation, commercial- and government-owned spent nuclear fuel (SNF) and high-level waste (HLW) will be loaded into casks and shipped to the Yucca Mountain repository. Materials will be transferred from the casks into a waste package (WP), sealed, and placed into the underground facility. The welding of lids onto the WP and the associated nondestructive examinations will be done in a closure cell by a welding and inspection system. Welding of the lids to the WP vessels will be performed using the cold wire, gas tungsten arc-welding process. Tack welds will be dressed (wire brushed and ground, as needed) before final welding. Weld passes will be wire brushed to remove fume particles and metal oxides after each pass. In-process vacuum cleaning will be used to remove debris remaining from weld brushing and grinding. Wire brushes and grinding tools used for 316 stainless steel will not be interchanged with wire brushes and grinding tools used for Alloy 22.

A spread ring assembly will retain the 316 stainless steel inner lid in position. The spread ring assembly will be tack-welded to the inner lid and 316 stainless steel inner vessel, and then seal-welded to the inner vessel by a two-pass seal weld. A purge port cap will be placed over the inner vessel purge port, tack-welded, and seal-welded in the same manner as for the inner lid-to-spread ring-to-inner vessel tack and seal welds. The Alloy 22 middle lid will be tack-welded and then multi-pass fillet-welded to the Alloy 22 shell. The outer lid will be tack-welded and narrow groove-welded to the shell.

At least two welding torches will be used to weld the inner, middle, and outer lids. These torches will be placed symmetrically about the respective weld joints during welding to ensure that residual stresses during welding are reasonably balanced across the lids. This is required to prevent movement of the lids during welding. This requirement does not apply for welding of the purge port cap to the inner lid. The welding equipment must be capable of and suitable for:

1. Tack and seal welding of 316 stainless steel WP components using the gas tungsten arc-welding process
2. Tack, fillet, and narrow groove welding of the Alloy 22 WP components using the gas tungsten arc-welding process

Specific welds that must be completed include:

1. Tack and seal welding of the spread ring assembly to the inner lid and the inner vessel
2. Tack and seal welding of the WP purge port cap to the inner lid
3. Tack and fillet welding of the WP middle lid to the outer corrosion barrier
4. Tack and narrow groove welding of the WP outer lid to the outer corrosion barrier

The welding equipment must be capable of and suitable for removing and repairing minor defective welds, as required.

Inspection methods for the WP include quantitative visual, eddy current, and ultrasonic inspection of lid seal and fillet and groove welds. Defects less than 0.063 in (1.5 mm) on all sides are acceptable. The equipment must be capable and suitable for performing the required inspections for the welds listed above. Specific inspections that must be completed are as follows:

1. Tack welds between the 316 stainless steel vessel, the spread rings, and the inner lid will be visually examined qualitatively. The two-pass seal welds between the 316 stainless steel vessel, the spread rings, and the inner lid will be visually inspected quantitatively.
2. The two-pass seal weld for the spread ring splice will be visually inspected quantitatively.
3. Tack welds between the purge port cap and the inner lid will be visually examined qualitatively. The two-pass seal welds between the purge port cap and the inner lid will be visually inspected quantitatively.
4. Tack welds between the Alloy 22 middle lid and the Alloy 22 shell will be visually examined qualitatively. The multi-pass fillet weld between the Alloy 22 middle lid and the Alloy 22 shell will be visually inspected quantitatively and eddy current inspected following completion.
5. Tack welds between the Alloy 22 outer lid and the Alloy 22 shell will be visually examined qualitatively. The multi-pass, narrow groove weld between the Alloy 22 outer lid and the Alloy 22 outer shell will be ultrasonically inspected on a pass-by-pass basis following each weld pass. It will be visually inspected quantitatively; eddy current inspected; and ultrasonically inspected following completion of welding. It will also be visually inspected quantitatively; eddy current inspected; and ultrasonically inspected following completion of stress mitigation.

In addition, the eddy current subsystem must be capable of inspecting the surface of weld repair cavities.

The welding and inspection system has to:

1. weld, inspect, and repair lids and purge port cap.
2. provide access for lids and other tools.
3. be maintainable and serviceable either in the glovebox; the closure maintenance area; or in another facility following decontamination, bagging, and removal from the closure cell.
4. be reliable.
5. be recoverable from off-normal events.

The primary risks that must be considered during the design of the system are:

1. Failure to close the WP
2. Failure of a subsystem during WP closure
   2.1. Requiring repair of a WP
   2.2. Requiring personnel entry into the closure cell
   2.3. Increasing the cycle time for closure of a WP
3. Low equipment reliability, requiring excessive maintenance or replacement

The order of priority in the specification of hardware will be as follows:

1. Commercial off-the-shelf hardware will be used when it can reasonably meet functional and operation requirements

2. Modified commercial hardware will be used when off-the-shelf hardware is not adequate

3. Custom hardware may be used when neither of the above two options are adequate

The subsystems will be modular. The primary reason for this is to provide an adequate effective design life of the system. The closure cell, in which the welding and inspection subsystem will be used, is expected to have a design lifetime of several decades. Some of the equipment in the closure cell, such as the overhead crane, can easily be specified to meet such a design lifetime. However, the welding and inspection subsystem is expected to become unreliable and require replacement several times over such a period. In addition, various subsystems are expected to become obsolete, with significantly more advanced technology available for potential replacement. Finally, it is recognized that the subsystems and components of the system will eventually fail; thus, it is important to make it easy to replace subsystems and components. An open-system architecture will be used to allow replacement or substitution of various subsystems and components without need for redesign of the total system. Although proprietary, single source technology may be used for portions of the total system, it is intended that such technologies will be employed at the subsystem or component level in a manner that such technologies can be replaced by alternative technologies. This will be done to ensure that long-term system viability will not be compromised by the actions or business fortunes of particular technology sources.

CONFIGURATION INFORMATION

Positioning requirements

Welding torches will need to be positioned laterally within about ±0.1 mm of the desired position over the weld joint during welding. The welding torch axis shall be positioned within about ±1 degree of the desired angular position. Such positioning could be done directly by robotic or seam-tracking mechanisms located on the welding end-effectors. In the latter case, the robotic mechanisms would need to position the welding end-effectors within about ±1 mm of the desired position. The welding torches will need to be positioned such that the electrode tips would be within about ±0.1 mm of the desired distance (in a direction parallel to the electrode centerline axis) from the weld joint or prior weld bead during welding to maintain a proper arc voltage. Such positioning could be done directly by the robotic mechanisms or by arc-voltage control mechanisms located on the welding end-effectors. In this latter case, the robotic mechanisms would need to position the welding end-effectors within about ±1 mm of the desired position.

Positioning requirements for the quantitative visual and eddy current inspection systems are expected to be nominally within about ±1 mm of their desired positions. Such positioning could be done directly by robotic mechanisms. The ultrasonic inspection systems will need to be positioned within about ±1 mm of their desired positions. These systems will also need to be pressed into direct contact with the WP. A suspension system may be designed for the ultrasonic and eddy current inspection systems to augment real-time force-feedback positioning control.

Motion control

The traditional concept of motion control involves mechanisms, motors, and control electronics. However, in this case, of concern are those process motions required to close a WP, as follows:

1. The inner lid of the WP is held in place structurally with a spread ring, but the spread ring must be sealed to both the inner lid and the inner vessel. First, a set of 16 tack welds will be made: eight between the spread ring and the inner lid, and eight between the spread ring and the inner vessel. Then, two single-pass seal welds will be made between the spread ring and the inner vessel (one on each side for 180 degrees of circumference, plus some overlap at each end); two single-pass seal welds will be made between the spread ring and the inner lid; and two single-pass seal welds will be made between the two ends of the spread ring.
2. The purge port cap will be welded to the inner lid. Four tack welds will be made between the purge port cap and the inner lid. Then, two single-pass seal welds will be made between the purge port cap and the inner lid.
3. The middle lid will be fillet-welded to the Alloy 22 shell. Several tack welds will be made between the middle lid and the Alloy 22 shell. Then, several weld passes will be made between the middle lid and the Alloy 22 shell on each half of the WP, with overlap of the weld pass ends.
4. The outer lid will be narrow groove-welded to the Alloy 22 shell. A set of eight tack welds will be made between the outer lid and the Alloy 22 shell. Then, the outer lid will be welded to the Alloy 22 shell with multiple weld passes between the outer lid and the Alloy 22 shell on each half of the WP, with overlap of the weld pass ends.
5. Weld starts and stops need to be overlapped to prevent starts and stops from stacking up over each other. In addition, robotic trajectories typically involve rapid motion from a home position to a point near the weld starting point; slower movement to the start point; controlled tracking motion to the weld end-point; slow retraction to a point near the stop point; and then rapid motion back to the home position.

The result of the above operations is that a set of about 1,560 unique 3-dimensional points must be identified in the robot workspace over the WPs solely for welding. These points define various start and end-points for trajectories of weld torch motion. Consequently, there is a relatively large motion-control problem; this problem becomes much larger when tool storage, inspection, dressing, and weld repair operations are included.

For a system comprising two large robots mounted to the closure cell floor (for welding alone), there would be a need to deal with 312 error-sensitive, 3-dimensional points to close the WP. All welding would be done along paths requiring multi-axis robot motion. However, for a system comprising two robots mounted to two moving carriages on a circular track, the need would be to deal with eight error-sensitive 3-dimensional points; 116 error-sensitive 2-dimensional points; and 158 error insensitive 1-dimensional points. Welding is performed along paths requiring single axis carriage motion, with the exception of the 3-dimensional path for seal-welding the end of the spread ring.

Equipment configuration

The basic configuration of the welding and inspection equipment will be that of a concentric track machine. The track will be mounted about 6 inches (15 cm) above the operating floor of the closure cell, concentrically to a large hole in the operating floor with a diameter of 108 in (2.7-m). It will allow the WP to be placed in position below the nominal center of the concentric track for welding and inspection.

Two carriages will be placed on the concentric track to move the robots around the WP during welding and inspection. Control, data, power, and other utilities cables (and hoses) will be run from the closure cell control area to these carriages. A cable management system will be provided to allow these cables to move as needed during motion of the carriages. The carriages will be capable of motions in excess of 180° around the concentric track. This will allow for overlap of both welding and inspection lengths.

The carriages will be provided with commercially derived 6-axis robotic arms that will allow the various welding and inspection end-effectors to be placed in position on the WP, as required. The robotic arms will have sufficient range of motion to allow them to move themselves and any attached end-effectors to a position that will allow lids to be placed on the WP without removing the carriages from the concentric track. The arms will have sufficient reach to allow them to be used to weld the WP purge port cap to the inner lid.

A set of four end-effectors will be provided on each carriage: One for welding; two for inspection; and one for repair of the WP.

These end-effectors will connect to the robotic arms by means of quick-release tool change connectors, and will be stored in a tool tray on the carriage when not in use. The tool tray will connect to the carriage by means of a quick-disconnect tool plate, and will incorporate a second quick disconnect on top to allow the remote handling system (RHS) to easily move the tool tray to a glovebox in the support area for servicing. Figure 11 presents the diagram of the concentric weld track, carriage, articulated arm, and weld end-effector.

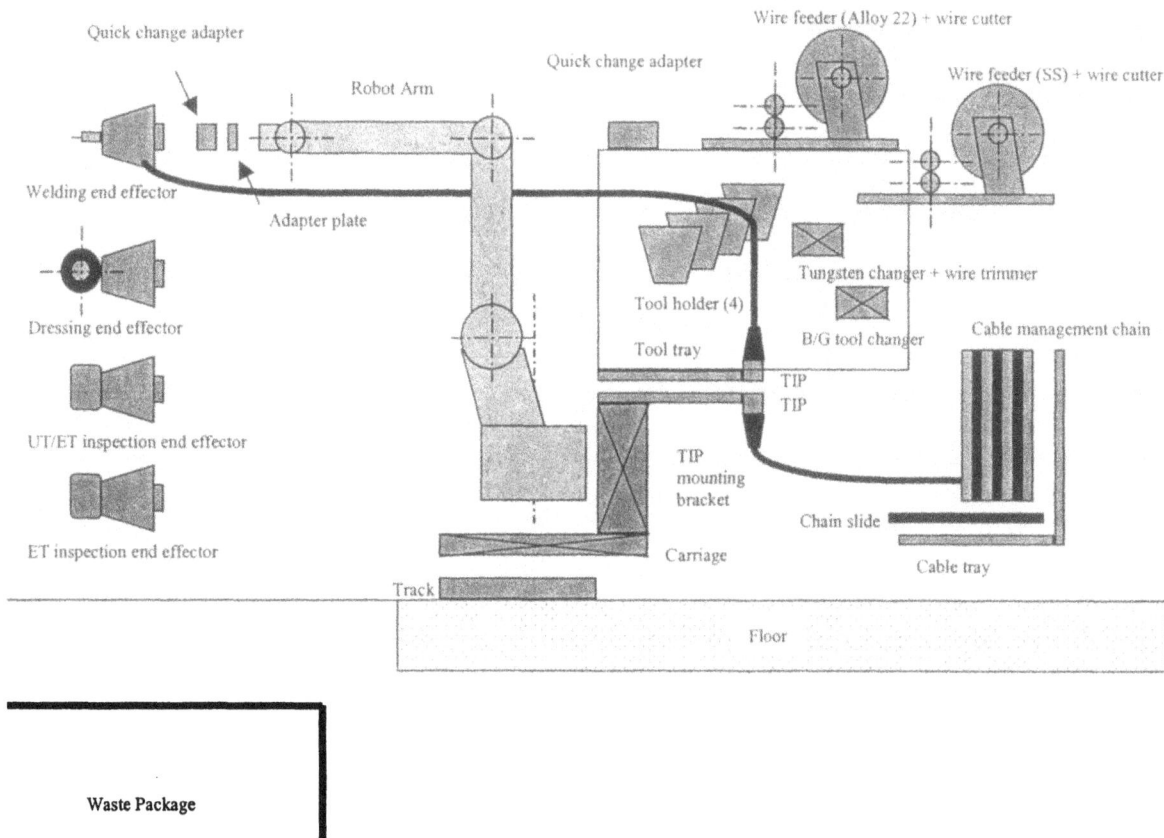

Fig. 11. Diagram showing concentric weld track, carriage, articulated arm, and weld end-effector (INEEL 2004b).

SYSTEM COMPONENTS

Concentric weld track

The concentric track will provide a guiding surface for the carriages. A ring gear will be mounted on the track to provide a means for carriage drive. The track will be mounted rigidly to the closure cell floor in a position close to floor height, and concentrically with the hole in the floor that provides access to the top of the WP.

Carriages

Two identical carriages will be used for circumferential motion of the welding and inspection systems above the top of the WP. Each carriage will incorporate a drive motor, enabling it to be driven about the concentric track. A pinion gear will be mounted to the output shaft of the motor, which will engage the ring gear mounted on the track. Ball or roller bearing metal wheels will be mounted to the carriages to guide their movement along the concentric track.

Robotic manipulator arms

Two nominally-identical 6-axis robotic arms will be used for welding; and quantitative and qualitative visual, eddy current, and ultrasonic inspection of the WP closure welds. One robot arm will be mounted to each of the two carriages. Modified commercial welding robots will be used if possible. The robots may actually have a 7th axis which would be optional for carriage motion control. Each robot should have: 1) sufficient reach to place the end-effectors in any position between the centerline of the inner lid and the outer rim of the upper lid weld joint; 2) a payload capacity of 30 to 40 kg (66 to 88 lb); and 3) weigh about 200 to 300 kg (440 to 660 lb).

Welding end-effectors

Two identical welding end-effectors (one per robot) will be used for welding both the 316 stainless steel and Alloy 22 components. This will require means to ensure that the appropriate filler wire and shielding gas is used for each weld; this way it will not be possible to weld 316 stainless steel using the filler wire and shielding gas intended for Alloy 22, and vice versa. Each welding end-effector will incorporate means for quantitative visual inspection of the weld passes, as required. These inspections will be performed using a seam-tracking sensor. This sensor, when scanned along the weld joint, will be capable of making 3-dimensional measurements of both the weld joint and weld surface profiles, and will be mounted ahead of the welding torch. It is possible to incorporate the seam-tracking sensor into one of the other end-effectors, and thus potentially reduce the radiation exposure it receives. However, the greatest risk is the failure to close the WP. By incorporating the seam-tracking sensor into the welding end-effector, it is possible to derive the most direct sensing of the weld joint configuration at the time of welding. Thus, this is the best choice to directly address seam-tracking needs during welding, and therefore reduce the probability of not successfully completing a closure weld.

A welding end-effector will incorporate a weld vision camera that will have suitable arc-light attenuation capabilities to provide video images of the weld pool during welding. The camera will be mounted in front of, and behind, the welding torch. Each welding end-effector will also incorporate video cameras to obtain video images of the weld bead leading edge and behind the welding torch.

The end-effector will incorporate a remotely-adjustable filler wire guide mechanism to ensure that the filler wire enters the weld pool in such a manner as to facilitate making a good weld. Each welding end-effector will incorporate a thermocouple or other temperature sensor capable of measuring the temperature of the base metal before each weld pass.

A welding end-effector will incorporate a mechanism to provide lateral motion of the weld torch normal to the welding direction. This mechanism will be used for seam-tracking and weld-torch oscillation during welding. Each welding end-effector will also incorporate a mechanism to provide vertical motion of the weld torch parallel to the axis of the torch. This mechanism will be used for automatic voltage control during welding and arc touch starting. Although motion in directions both lateral and parallel to the axis of the torch may be made using the robotic arms, this approach is preferred at this point in the system design for the following reasons:

1. Seam-tracking and automatic voltage control are the highest bandwidth motion control functions required for closure of a WP
2. Relative high bandwidth motion of small hardware components (e.g., 5 lb $\approx$ 2.2 kg) is easier and more reliable than similar motion of large hardware components (e.g., 500 lb $\approx$ 226 kg)
3. Simultaneous seam-tracking; torch oscillation; automatic voltage control; and arc touch starting are at the leading edge of robotic technology today

Dressing end-effector

Two weld dressing end-effectors (one per robot) will be used for wire brushing and grinding of weld joints and welds. Unique brushing and grinding tools will be provided for 316 stainless steel and Alloy 22 welds. Means will be provided to ensure that the incorrect tool is not used, to prevent contamination of the welds.

Ultrasonic and eddy current inspection end-effector

Two ultrasonic inspection end-effectors (one per robot) will be used for volumetric inspections. These may employ membranes on the contact surfaces of the transducers to allow small quantities of water to be used to achieve a low-impedance contact between the transducers and the WP. The transducers will likely use phased array technology. Several transducers may be used to provide a sufficient number of sound paths to adequately inspect the volume of the outer lid to the Alloy 22 shell closure weld. Design precautions will be taken to ensure that only small quantities of water could be lost during an off-normal event. The ultrasonic transducers will not be placed over the WP until both the inner and middle lid welds are completed. Consequently, it is not expected to have a significant risk from a criticality standpoint posed by using small quantities of water for ultrasonic coupling. Each of these end-effectors may also incorporate an eddy current transducer array to facilitate surface inspection of the outer lid to the Alloy 22 shell closure weld.

Eddy current inspection end-effector

Two eddy current inspection end-effectors (one per robot) will be used for surface inspection of the WP. Each end-effector may incorporate two transducer probes to facilitate surface inspection of the middle lid fillet weld and weld repair cavities.

Tool trays

Each carriage will incorporate a remotely-removable tool tray. The tool tray will be used for storing end-effectors, dressing tools, and electrodes. It will also hold the filler wire feeders; filler wire spools; gas cup/electrode stickout adjustment devices; electrode changer; dressing tool changers; and possibly—one or more video cameras. There will be a dedicated storage mount (holster) for each of the end-effectors on the tool tray. The lower end of the holsters for the welding; eddy current testing (ET) inspection; and ultrasonic testing (UT)/ET inspection end-effectors will incorporate calibration fixtures to be used to ensure proper operation of the various weld inspection devices. There will be a device on the tool tray that will facilitate changing tungsten welding electrodes. This device will incorporate storage for several new electrodes and separate storage for used electrodes. Means will also be provided for adjusting the electrode stickout. There will be means for changing weld dressing tools, including storage for several grinding wheels and wire brushes. Separate filler wire feeders will be mounted on the tool tray for 316 stainless steel and Alloy 22 welds. The wire feeders will incorporate sensors to measure wire speed to make measurements of the amount of filler wire consumed during welding. The tool tray will be mounted to the carriage by means of a remotely-operable quick-disconnect tool plate. Most of the electrical conductors and utilities to the end-effectors will pass through quick-disconnect plugs on the tool plate. A second remotely-operable quick disconnect will be mounted to the upper part of the tool tray to provide a means for the RHS to move the tool tray to the glovebox for service and maintenance.

Cable management system

A cable management system will be used to control the motion of hoses and electrical cables running between the closure cell wall and the carriages. A cable management chain will support the various cables and hoses, operating in a semi-circular tray outside the circumferential track. Several approaches have been considered for handling the various cables required for operating the welding and inspection system, as follows:

1. **Switching Network:** Using a switching network actually reduces the number of cables in the closure cell environment. This also reduces the set of cables passing through the closure cell wall to some minimum number needed for operating the robotic system and at least one end-effector. However, manual switching would not be reliable enough, and automatic switching would require a large switching system due to the number of cables needed in the system.

2. **Electronics in the Closure Cell:** Placing almost all the control and sensing electronics in the closure cell would reduce the number of cables passing through the closure cell wall to those required for welding power, instrument power, and communications. However, this would require closure cell space for the equipment; require radiation hardening of the electronics; and would significantly increase the difficulty of maintenance and repair of that equipment.

3. **Dedicated End-Effector Cables:** Dedicated end-effector cables with the control and sensing electronics outside the closure cell will increase the number of cables required to their maximum number.

Cables would most likely fail if they are severely flexed and if they are exposed to high levels of cumulative radiation exposure. Thus, cables from the carriage tool plates to the end-effectors will be more likely to fail than the cables from the closure cell wall to the carriage tool plates. The cables from the tool plates to the end-effectors can be readily replaced in the glovebox if they are dedicated to each end-effector. However, the cables that run along the robot arms would require personnel entry into the closure cell to replace them. Those cables would also experience considerably more severe flexing than the cables that run directly to the end-effectors, and would thus be more likely to fail.

Motors and position sensors

Motion control systems generally use either servomotors or stepping motors. Stepping motors are generally low-cost, reliable, and perform relatively well in radiation fields. However, they are open-loop-controlled devices and are generally used for low-speed, constant load applications. Servomotors are closed-loop-controlled devices generally used for higher speed and variable load applications. Because they are closed-loop controlled, servomotors require some form of position or speed-feedback sensing. Normally, for welding applications, optical encoders are used to provide the feedback sensing because they are relatively insensitive to the electrical noise generated during welding. However, radiation-hardened optical encoders are apparently not commercially available. For radiation applications, resolvers are normally used for feedback-sensing with servomotors. Fortunately, both radiation-hardened servomotors and resolvers are commercially available. Unfortunately, resolvers are not normally used for welding applications, as they are sensitive to electrical noise.

The solution to the above situation could be as simple as using stepping motors for all functions requiring motor drive in the radiation field. However, detection of motor stalls is an issue for stepping motors. Normally, encoders are used for stall detection, but recently encoderless stall detection has become available. But encoderless stall detection apparently does not work for motor stalls occurring during startup of the motor (i.e., at zero speed). In addition, commercial robots typically use encoder-equipped servomotors, which cannot be easily replaced with stepping motors without extensive changes to the control electronics and software.

Experiments indicate that the gas tungsten arc-welding process, using touch starting of the arc, does not produce enough electrical noise to present a problem when using resolvers for position sensing. This will allow INEEL to either employ radiation-hardened robots that already incorporate resolvers (which are commercially available), or to retrofit resolvers into encoder-equipped robots using resolver-to-encoder converters, which are also commercially available.

Welding power supply

The welding power supply will be a commercial grade unit incorporating pulsed current controls. The power supply will incorporate an analog or Ethernet input interface, allowing pulsed current waveform parameters to be specified from an external source, including an independent PC-type computer.

Tool/electrode changers

Means will be provided to replace used tungsten electrodes, wire brushes, and grinding wheels while the welding equipment is in position on the concentric track machine.

Electrical power

In the operating gallery, 460–480 V alternate current, 3-phase power will be required for the two welding power supplies in addition to 110 V alternate current for general use. In the closure cell, 110 V alternate current will be provided. Equipment requiring other electrical power (e.g., 5 V direct current for video cameras) will obtain that power from dedicated power supplies operating from one of the two main sources.

SYSTEM CONTROL FEATURES AND OPERATIONS

Control system

A commercial off-the-shelf control system will be used to monitor and control the various functions of the welding and inspection system.

User interface

A human/machine interface (HMI) will allow human operators to operate the welding and inspection system. The HMI will incorporate a graphical user interface (GUI) displayed on one or more computer screens, with additional control devices, such as joysticks, mice, track ball, and possibly other devices.

Control computer

A PC-type computer will be used to manage the GUI; execute high-level motion control algorithms; and control the welding process. Application software will be developed in various programming environments and languages using LabView; C; C++; Java; or some combination thereof. Ethernet and TCP/IP will be used for systems-level communications. Additional analog and digital communications will be used for low-level control, as required.

Sensors and control equipment

Sensors will be used for various functions, including weld seam-tracking; automatic voltage control of the welding process; WP temperature measurements; and video images of the WP and welding process. Normal operation of equipment will be in a semi-autonomous mode. Procedures will be pre-loaded into the control computers and designated for execution by the operator. Equipment will incorporate hard-wired emergency stop switches that will turn-off power to system actuators (including welding power supplies; robotic arms; carriages; wire feeders; and other energized devices that could compromise personnel, equipment, and WP safety) from a single-point operator's station. Computer-controlled soft-stop procedures will also be incorporated that will cause all system-active components to assume a predesignated condition selected to ensure personnel, equipment, and WP safety. Components located in the closure cell will be designed as easily-handled modules suitable for maintenance in a glovebox environment, to the extent possible. Components will be provided with suitable features, such as quick-release connections for cables and hoses, and release handles or actuators to allow remote removal of components for larger systems to support remote maintenance.

DESIGN REQUIREMENTS

Detailed design requirements, including the corresponding performance measures and bases, are listed in Table 22.

Table 22. Welding and Inspection System Design Requirements (INEEL 2004b).

Number	Requirement	Performance Measure	Basis
WI-3-1	The welding and inspection system shall have adequate functionality to perform the required welds and inspections.	Section 1 of TFR-283 (INEEL 2004b) and Section 1.1 of TFR-282 (INEEL 2004a).	EDF-4227 (INEEL 2004g) See TFR-282, F&OR Requirements 1.1.2.3.2-1 (NFD) and 1.1.2.3-3. (NFD)
WI-3-2	Where practical, tools and equipment shall be designed to accommodate the different sizes of waste package and perform multiple tasks to minimize the number of tools.	Demonstration of the extent to which the number of tools has been minimized.	TFR-282, section 4, Design Guideline 15
WI-3-3	The welding and inspection system shall weld and inspect lid and purge port cap closure welds.	Demonstration of the welding and inspection system welding and inspecting the lids and cap with remote welding and inspection equipment using qualified procedures.	10 CFR 63.111 (US NRC 2003c) 10 CFR 63.113(b)-(c) (US NRC 2003d)
WI-3-4	All welds will be made using the cold-wire gas tungsten arc welding process.	Demonstration that the welding and inspection system design has incorporated the cold-wire gas tungsten arc welding process.	Canning (2004).
WI-3-5	Welding shall be done using two welding end effectors located nominally 180 degrees apart, capable of concurrent and independent operation from the control consoles.	Demonstration that the welding and inspection system design has incorporated two welding end effectors located nominally 180 degrees apart, capable of concurrent and independent operation from the control consoles.	Design discussions with BSC held at INEEL Nov. 18-19, 2002. EDF-4227 (INEEL 2004g). Good engineering practice to eliminate lid movement during welding due to weld-induced stresses.
WI-3-6	The welding and inspection system shall be capable of using different shielding gas compositions, as needed, for 316 stainless steel and Alloy 22 welding.	Demonstration that the welding and inspection system was designed to incorporate two shielding gas systems for the different shielding gas compositions, as needed, for 316 stainless steel and Alloy 22 welding.	Good engineering practice.
WI-3-7	The welding and inspection system shall be capable of performing welding and inspection activities in the closure cell environment.	Demonstration welds will be made in demonstration facility duplicating the closure cell environment.	See TFR-282 (INEEL 2004a) Section 1.1.
WI-3-8	The welding and inspection system shall have means of providing the motion and motion control necessary for operating the in-cell portions of the system.	Motion and motion control will be demonstrated in the demonstration facility.	EDF-4227 (INEEL 2004g)

Table 22. (cont'd).

Number	Requirement	Performance Measure	Basis
WI-3-9	The welding and inspection system shall incorporate sufficient range of motion to be able to position the end effectors at the proper locations to carry out their functions.	Demonstration of ability to position the various end effectors appropriately over the waste package weld joints for the full range of waste package sizes.	Good engineering practice.
WI-3-10	The welding and inspection system shall have capabilities for integration with other closure cell systems.	The performance of the interface shall have no interface error messages during the fullscale waste package processing validation demonstration.	Good engineering practice.
WI-3-11	The welding and inspection system shall have means of managing the various cables and hoses necessary for operating the in-cell portions of the system.	Cable management will be demonstrated in the demonstration facility.	Good engineering practice.
WI-3-12	The welding and inspection system shall be designed for safe operation in accordance with any applicable laser safety requirements.	Demonstration that the nominal hazard distance for the laser is contained within the closure cell during operation or within an optically shielded glove box during maintenance.	See TFR-282 (INEEL 2004a), Section 4, Design Guideline 6
WI-3-13	Online operating efficiency shall be 60%, and welding operations shall be completed within 70 hours (44 hr for optimum, 100%, efficiency) of the WP being secured in the closure cell. The welding and inspection system will be designed to meet productivity and quality requirements in the closure cell.	Demonstration that online operating efficiency is 60%, and welding operations can be completed within 70 hours (44 hr for optimum, 100%, efficiency) of the WP being secured in the closure cell.	Design discussions with BSC held at the INEEL Nov. 18-19, 2002. See TFR-282 (INEEL 2004a) WPCS TR-3.1.2.
WI-3-14	Equipment containing liquids shall be designed to reasonably preclude leakage of those fluids into the cell. Small amounts of free liquids may be released into the WPCA directly associated with ultrasonic inspection of the narrow-groove closure weld between the outer lid and the alloy 22 shell.	Demonstrate by design review that equipment containing liquids are designed to reasonably preclude leakage of those fluids into the cell. Small amounts of free liquids may be released into the WPCA directly associated with ultrasonic inspection of the narrow-groove closure weld between the outer lid and the alloy 22 shell.	INEEL Waste Package Closure System Technical Team. See TFR-282 (INEEL 2004a), WPCS TR-3.6.4-5
WI-3-15	The welding and inspection system shall be capable of cleaning the weld joint and adjacent areas by wire brushing.	Demonstration that the welding and inspection system incorporates a dressing end effector capable of wire brushing weld joints and closely adjacent areas.	INEEL Waste Package Closure System Technical Team. Good engineering practice.

Table 22. (cont'd).

Number	Requirement	Performance Measure	Basis
WI-3-16	The welding and inspection system shall be capable of welding and repairing defective welds.	The welding and inspection system shall successfully demonstrate in a process validation demonstration the capability of grinding and weld repairing an indication on a test coupon.	BSC design decision (Lundin 2002). See TFR-282 (INEEL 2004a) F&OR Requirement 1.1.2.3.2-1 (NFD)
WI-3-17	The welding and inspection system shall inspect repair cavities and associated repair welds for nonconformities. Nondestructive examination techniques shall be capable of detecting 0.063 in. (1.5 mm) or greater flaws.	Demonstration that the nondestructive examination techniques are capable of detecting 0.063 in. or greater flaws.	EDF-4214 (Kunerth 2004) PRD-002/T-012 (Curry and Loros 2002a) PRD-002/T-014 (Curry and Loros 2002b)
WI-3-18	The welding and inspection system shall have the necessary functionality needed for maintenance.	The welding and inspection system subsystems will be designed for maintenance in the glove box or cell maintenance areas, or they may be bagged out for maintenance in some other (unspecified) location.	INEEL Waste Package Closure System Technical Team. See TFR-282 (INEEL 2004a) F&OR Requirement 1.1.2.3-2a.
WI-3-19	All welding equipment located in the closure cell will be designed for recovery by the maintenance system.	Demonstration that all welding and inspection system equipment located in the closure cell are designed for recovery by the RHS or overhead crane.	INEEL Waste Package Closure System Technical Team. Good engineering practice.
WI-3-20	The welding and inspection system components used in the closure cell requiring frequent maintenance or servicing will be designed for glove box maintenance or replacement.	Demonstration that welding and inspection system equipment requiring frequent maintenance or servicing is designed for transport to the glove box by the RHS for maintenance, servicing, or replacement.	INEEL Waste Package Closure System Technical Team. Good engineering practice.
WI-3-21	The welding and inspection system components not designed for glovebox maintenance, servicing, or replacement will be designed for remote or in-cell maintenance.	Demonstration that welding and inspection system equipment not designed for glove box maintenance, servicing, or replacement is designed to be bagged out for remote maintenance or maintained in the closure cell maintenance area.	Good engineering practice.
WI-3-22	For decontamination reasons, tooling and equipment exposed surface finishes of 32 micro-inch (0.8 mm) or better shall be considered where appropriate.	Demonstration that exposed surface finishes of 32 micro-inch or better are used on equipment where appropriate.	Good engineering practice.

Table 22. (cont'd).

Number	Requirement	Performance Measure	Basis
WI-3-23	The welding and inspection system shall have means of controlling the system remotely from the operating gallery.	Demonstration that the welding and inspection system is capable of operating remotely in the closure cell under the auspices of computer control, with human guidance, or a combination of both.	See TFR-282 F&OR Requirement Number 1.1.2.3-5 (NFD).
WI-3-24	The welding and inspection system shall be capable of acquiring data, as applicable, to support waste package final documentation.	Demonstration that the welding and inspection system is capable of acquiring data to support WP final documentation, as applicable.	See TFR-282 (INEEL 2004a) Section 1.1.
WI-3-25	For decommissioning, it shall be considered that the welding and inspection system equipment within the WPCS be designed for removal without demolition.	Demonstration that the welding and inspection system equipment within the WPCS is designed with small enough subsystems or components to allow the entire system to be removed from the closure cell through the glove box.	Good engineering practice.
WI-3-26	Unnecessary tool contact on and near the weld preps prior to welding shall be avoided to protect the weld zone from damage and impurities.	Demonstration that during normal operation only the ultrasonic inspection transducers and the dressing tools will contact the WP.	Good engineering practice.
WI-3-27	Equipment and tooling shall be designed for long-term use and have provisions for easy replacement and upgrade of worn or damaged parts.	Demonstration that the welding and inspection system is designed in a modular form to allow for relatively easy replacement and upgrade of worn or damaged components.	Good engineering practice.
WI-3-28	The welding and inspection system shall incorporate calibration blocks for VT, ET, and UT to validate operation of the inspection system.	Demonstration that the VT, ET, and UT calibration blocks are located on the tool trays for validation of operation.	Good engineering practice.
WI-3-29	The welding and inspection system calibration blocks shall be in the closure cell or maintenance area for ready inspection process calibration.	Demonstration that the VT, ET, and UT calibration blocks are located in the glove box for inspection process calibration.	Good engineering practice.
WI-3-30	Welding and inspection system carriages will be administratively controlled so as to come no closer than 100 degrees apart.	Demonstration that the welding and inspection system controls coordinate carriage motion to ensure that the carriages do not come closer than 100 degrees apart.	Good engineering practice to avoid robot collisions, to and ensure welding induced stresses are reasonably balanced across WP.

Table 22. (cont'd).

Number	Requirement	Performance Measure	Basis
WI-3-31	Welding torches shall be positioned laterally within about ±0.1 mm of the desired position over the weld joint during welding.	Demonstration that the welding torches are positioned laterally within about ±0.1 mm of the desired position over the weld joint during welding.	Good engineering practice.
WI-3-32	The welding torch axis shall be positioned within ±1 degree of the desired angular position.	Demonstration that the welding torch axis is positioned within ±1 degree of the desired angular position.	Good engineering practice.
WI-3-33	If positioned by robotic mechanisms or by seam tracking mechanisms located on the welding end effectors, the mechanisms shall position the welding end effectors to within about ±1 mm of the desired position.	Demonstration that if positioned by robotic mechanisms or by seam tracking mechanisms located on the welding end effectors, the mechanisms position the welding end effectors to within about ±1 mm of the desired position.	Good engineering practice.
WI-3-34	Welding torches shall be positioned such that the electrode tips are within about ±0.1 mm of the desired distance (in a direction parallel to the electrode centerline axis) from the weld joint or prior weld bead during welding to maintain proper arc voltage.	Demonstration that the welding torches are positioned such that the electrode tips are within about ±0.1 mm of the desired distance (in a direction parallel to the electrode centerline axis) from the weld joint or prior weld bead during welding to maintain proper arc voltage.	Good engineering practice.
WI-3-35	If positioned by robotic mechanisms or by arc voltage control mechanisms located on the welding end effectors, the robotic mechanisms shall position the welding end effectors within about ±1 mm of the desired position.	Demonstration that if positioned by robotic mechanisms or by arc voltage control mechanisms located on the welding end effectors, the mechanisms position the welding end effectors within about ±1 mm of the desired position.	Good engineering practice.
WI-3-36	Positioning requirements for the quantitative visual and eddy current inspection systems shall be nominally within about ±1 mm of their desired positions.	Demonstration that the positioning requirements for the quantitative visual and eddy current inspection systems are nominally within about ±1 mm of their desired positions.	Good engineering practice.
WI-3-37	The ultrasonic inspection systems shall be positioned within about ±1 mm of their desired positions.	Demonstration that the ultrasonic inspection systems are positioned within about ±1 mm of their desired positions.	Good engineering practice.
WI-3-38	The ultrasonic inspection systems shall be pressed into direct contact with the WP.	Demonstration that the ultrasonic inspection systems are pressed into direct contact with the WP.	Good engineering practice.

SAFETY SYSTEMS

INTRODUCTION

The following material is derived from the TFR-295, Component Design Description: WPCS Safety System (INEEL 2004d) document.

The safety system for the Waste Package Closure System (WPCS) is designed to protect equipment from damage and personnel from injury. The WPCS safety system features a class of programmable logic controllers (PLCs) specifically designed for use in safety-critical applications. The objective of TFR-295 (INEEL 2004d) is to inform the equipment designers of WPCS safety system capabilities. Programmable safety systems designed to replace older safety relay technologies are less prone to "nuisance trips"; provide diagnostic information when trips occur; and cost less to install and maintain. The PLCs for safety applications are manufactured by several commercial vendors and are readily available. These modular units incorporate multiple independent processors that separately monitor inputs to determine their validity. If a trip occurs, these processors automatically compare data to determine whether the trip is valid. This tends to eliminate nuisance trips or false alarms that occur with safety relay systems. Programmable safety systems can also monitor input switch/relay circuits and cabling, and can be programmed to alarm if the circuit opens or shorts. All programmable safety PLCs provide emergency stops, enabling devices, and safeguard or protective devices in accordance with the International Standards Organization and American National Standards Institute standards.

DESIGN DESCRIPTION

The WPCS safety system can be programmed to protect specific machines, zones, or areas; and can be programmed to output either to a specific piece of equipment or to the whole system. The safety system comprises three processors manufactured by three different vendors, but designed to work in concert. They constitute a voting scheme that helps eliminate false triggers and ensures accurate alarm sensing. The system can be designed to monitor continuity through alarm switches and interlocks to help ensure that the systems are in order.

Figure 12 depicts an expandable, modular rack-mount model manufactured by PILZ. A network referred to as a *Safetybus* allows expansion of the system by adding additional input/output modules and additional racks of modules, if necessary. The system communicates with the supervisory control system by Ethernet [Transmission Control Protocol/Internet Protocol (TCP/IP)], reporting the status of interlocks, alarms, or the system in general to the closure cell control system.

A variety of input/output modules are available for convenient interface with most sensors, switches, and alarms. Some modules superimpose a pulse train signal on the alarm switch wiring and monitor the presence of the signal at an input on the module. If the pulse train signal is not detected, the module notifies the central processing unit, which actuates a programmed response.

Fig. 12. PILZ modular safety system components (INEEL 2004d).

Boundaries and interfaces

The safety system interfaces with equipment throughout the closure cell area. The following list identifies specific equipment interfaces that are already defined:

1. Supervisory control system
2. Remote handling system
3. Welding systems
4. Closure cell crane
5. Master/slave manipulator lockup
6. Glovebox interlock controls

Physical location, layout, and principles of operation

The safety system will be located in the equipment racks in the operating gallery. The architecture is a modular design that can be easily expanded. A standard software protocol will be used for communicating with the supervisory control system by Ethernet (TCP/IP).

System reliability and control features

The safety system is designed with redundant processors to preclude failures. If properly designed, interlock switches, cabling, and E-stop switches can be monitored for cable or switch failures, and alert the control system of these conditions. The system will have all appropriate testing certifications for safe operation, such as the Underwriters Laboratories (UL) listing or other nationally-recognized testing laboratories.

Operations

The safety system, which will be based in the closure operating gallery, will not be exposed to extreme temperatures or a radiation environment. The safety system will maintain all interlocks; emergency shutdown capabilities; and a network connection to the supervisory control system. It will be a triple voting system, certified by the appropriate evaluation laboratories (UL, Canadian Standard Association, etc.). If an interlock fails or an emergency stop occurs, the safety system will prevent operation of the equipment (safe standdown and lock), and report the occurrence to the supervisory control system (which has the capability of communicating to the digital control and management information system [DCMIS]). Once a shutdown has occurred, an operator will be required to follow a startup procedure to bring the equipment back on line. Inherent to the safety system design is constant monitoring and verification of the interlock/ emergency switch circuits. Password authorization is required to modify the safety system software. A separate computer with an RS-485 interface capability and the appropriate development software is required to develop or modify the program. The Ethernet (TCP/IP) interface does not allow program access. The control switches, such as "Power" and "Operate," will be protected by a locked panel or by key switches to preclude unauthorized disabling of the system.

Testing and maintenance

The complete safety system, including all associated hardware, will be tested and verified according to an approved test plan and procedure.

DESIGN REQUIREMENTS

The design requirements are derived from TFR-282 (INEEL 2004a) and best engineering practices. Table 23 lists the detailed requirements for the WPCS Safety Systems together with the corresponding performance measures, bases, and verification methods.

Table 23. Design Requirements (INEEL 2004d) for the WPCS Safety Systems as derived from TFR-282 (INEEL 2004a).

Requirement Number	Requirement	Performance Measure	Basis	Verification Methods
SS-1	Emergency shutdown switches will be installed at strategic points in the operating gallery and support area.	E-switches will be shown on the equipment layout drawing (Drawing 624835 [Housely 2003a]) and functionally verified during the system SO test.	TR-3.6.7-3 TR-3.6.12-4	Test and observation
SS-2	Safety interlocks will be installed on equipment in the closure cell and glovebox to preclude personnel injury and equipment damage.	Functional verification will be performed during SO tests.	TR-3.6.7-3 TR-3.6.12-5	Test
SS-3	Safety system status will be available to the supervisory control system and the DCMIS by Ethernet (TCP/IP) network.	Verification will be performed during SO tests.	TR-3.6.2-3	Demonstration
SS-4	Local interlock alarms and warnings will be available in designated areas in the closure cell.	Local alarms will be shown on equipment layout drawing (Drawing 624835 [Housely 2003a]) and functionally verified during system SO test.	TR-3.6.7-3	Observation

Table 23. (cont'd).

Requirement Number	Requirement	Performance Measure	Basis	Verification Methods
SS-5	Safety system control electronics will be located in the operating gallery.	The equipment will be located in a NEMA 4 rated wallmount enclosure with controlled access capability.	TR-3.6.7-4 TR-3.6.9-2	Observation
SS-6	The safety system control software will be controlled in accordance with the Software Quality Assurance Plan for the project.	Safety system software will follow project guidelines for software module documentation and control.	TR-3.6.8-1	Observation
SS-7	The safety system will be tested to ensure operation and compatibility with associated systems.	Integration verification during system SO test.	TR-3.7.1-1	Demonstration
SS-8	"Mean time between failures" and "mean time to repair" information will be obtained from the vendor.	MTBF and MTTR information will be included in the O&M manual.	TR-3.7.2-2	Observation
SS-9	The safety system will be modular to provide easy maintenance and repair and facilitate troubleshooting.	PILZ PSS3000 series programmable safety system will be used. This consists of plug-in modules and built in diagnostics.	TR Design Guideline 8	Demonstration and observation
SS-10	The safety system will be expandable to allow additional functions to be added.	The PILZ PSS3000 system allows expansion through the Safetybus system.	Best engineering practice to provide for facility interfaces as necessary.	Observation
SS-11	The safety system requires 120 V alternate current, 20-A power.	Power requirements are identified on Drawing 625198, 3-B (Housely 2003b).	TR-3.6.5-3	Observation

Boundary and interface technical requirements and bases

The safety system interacts with the supervisory control system and the digital control and management information system (DCMIS) by way of a data communications network (Ethernet using TCP/IP protocol). Each subsystem with interlock requirements will provide an interface compatible with the circuit monitoring capabilities of the safety system.

Utility systems

The facility will provide 120 V alternate current at 20 A power for operation of the safety system.

Testing

A system qualification test for design verification of the system control requirements will be performed before onsite system assembly and system operability (SO) testing. The qualification test will include as much of the system equipment as reasonably possible. An SO test will be performed on the subsystem before delivery and customer turnover. All or any other parts of the subsystem may be tested at this time. An integrated test of the complete system will take place on site before customer turnover. Control hardware will be maintained outside the closure cell using typical maintenance procedures.

Special technical requirements and bases

The control hardware will be located in the operating gallery, with remote stations in other areas as needed. The radiation hazards are not expected to be significant because the level of radiation in the operating gallery is designed to be less than 0.25 mrem/h. The safety system will be used to monitor and mitigate hazards resulting from electrical equipment operation or failures. It will be UL listed, and will be installed according to current state and local codes and the National Electrical Code (NFPA 2002a). Applicable standards relating to industrial machinery, like Electrical Standard for Industrial Machinery (NFPA 2002b), will be followed in programming the safety system. The safety system does not replace the need for lockout/tagout of equipment during maintenance or repair.

CONTROL AND DATA MANAGEMENT

INTRODUCTION

The following material is derived from the TFR-300, Component Design Description: WPCS Control and Data Management System (INEEL 2004e) document.

The Waste Package Closure System (WPCS) Control and Data Management System (C&DMS) has three levels of control, as follows:

1. The first level is supervisory, designed to administrate the operation of the subsystems under its jurisdiction
2. The second is process control, which directly allows operator interface to the third level
3. The third level is operations, which enables operator command and process control of real-time control modules for performing the weld-related operations

The architecture is modular and uses standard, commercially-available software packages to interface with the various hardware and software components in the WPCS. All three control levels will require custom software written to industry-standard protocols. The following facility-level interface requirements must be met in order for the C&DMS objectives to be accomplished:

1. Facility Operations will implement a digital control and management information system (DCMIS) at the facility level to interface with the C&DMS. The interface will enable transfer of data and video records between the two systems.
2. The spent nuclear fuel/high-level waste (SNF/HLW) transfer system will control the movement of the waste package (WP) transporter.
3. The DCMIS will control the ground floor airlock doors that isolate the WP beneath the closure cell, and will be responsible for all sensing requirements associated with this function.
4. The SNF/HLW transfer system will position the WP into each closure cell process opening, and will be responsible for all sensing requirements associated with this function.
5. The WP will be positioned in each closure cell process opening within a tolerance of ±2.0 in (±5 cm) in the X, Y, and Z planes, and level to within ±0.5 in (±12 mm).

The DCMIS will export weld process control data to the C&DMS, and will import weld quality, inspection, and verification data from the C&DMS. The DCMIS will also export an electronic document to the C&DMS for the WP that provides WP identification, and other essential data relative to the WP closure process. This electronic document is modified as the closure process progresses to reflect the changing status of the WP, and is returned to the DCMIS at completion of the process.

The TFR-300 (INEEL 2004e) document defines the design requirements and descriptions for the closure cell/operations gallery power and controls interface; support area/glovebox power and controls interface; operations gallery/support area controls interface; control electronics equipment locations; control software architecture; control software communications protocol to the hardware device control modules (HDCMs); software configuration management; database management; and DCMIS interface.

Figure 13 shows the control system architecture. The supervisory control station (SCS) will interface with the DCMIS and, when required, will send the operator workstations (OWs) facility information and alarms.

The SCS oversees operation of each of the closure cell OWs. This oversight requires a control scheme to collect and distribute the status of all control operations, and to provide permissives to each system to start or continue its operations. A supervisory control and data acquisition (SCADA) software package is used to help facilitate the communications between modules and provide easier maintenance and upgrades. Each workstation provides the means to control one or more of the HDCMs. Each of these modules performs one of the major functions shown at the bottom of Fig. 13. The HDCMs are the electronic systems that control the various hardware devices that make up the WPCS.

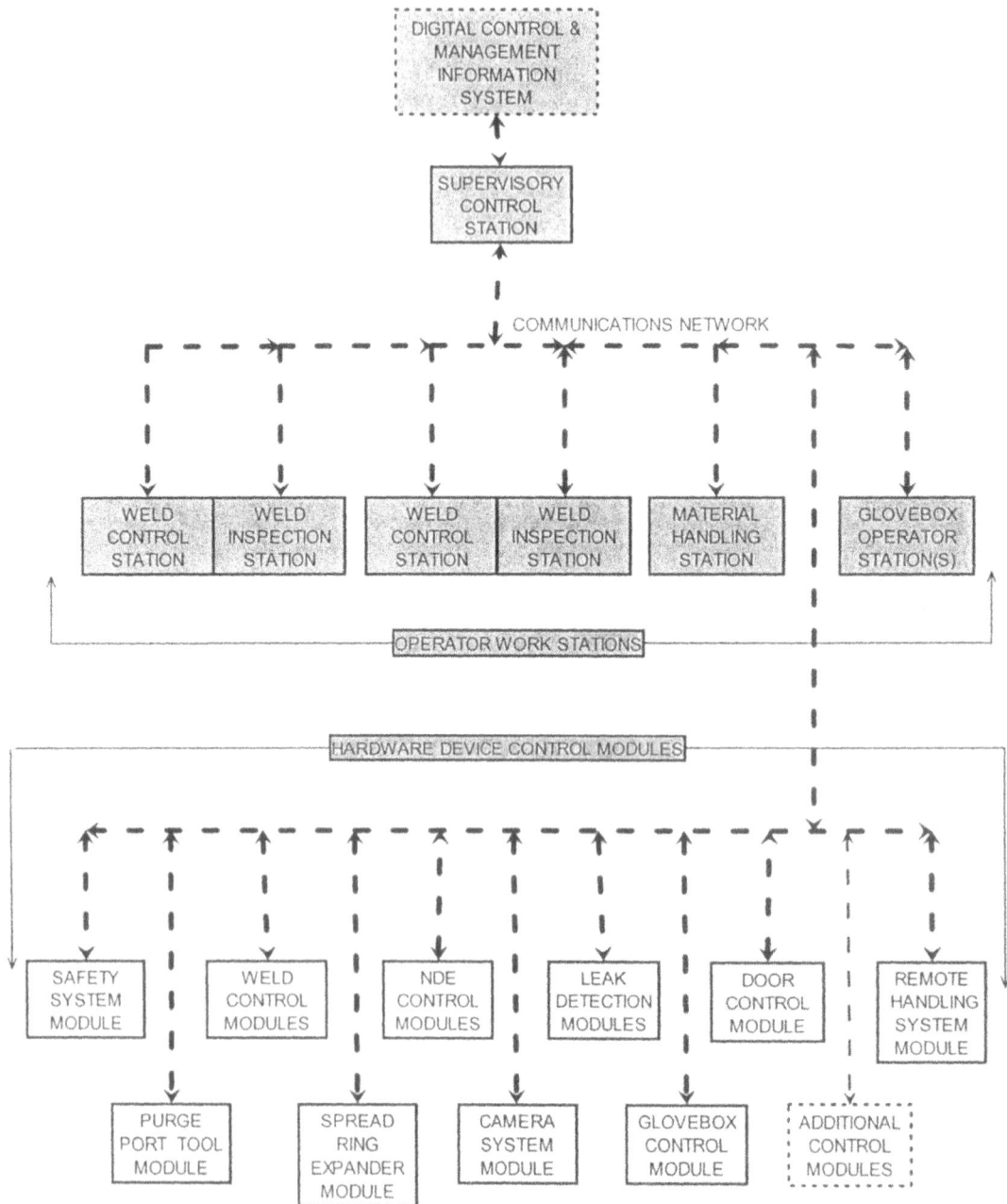

Fig. 13. Control system architecture (INEEL 2004e).

CLOSURE CELL/OPERATING GALLERY POWER AND CONTROL INTERFACE

The interfaces between the closure cell and the operating gallery include the following:

1. Module power and signal cabling through connector enclosures mounted inside and outside the cell.
2. Module signal cabling/utilities through K-plug connections. K-plug is a term used in industry to define a specially-designed plug placed in the closure cell wall to provide a more direct routing of cables/tubing to the operating gallery.
3. Electrical power interface for control electronics within the operating gallery.
4. Safety system interface.

Hardware device control modules

Hardware device control modules control the operation of hardware device modules (HDMs). Multiple HDMs may be controlled from one HDCM. Typically, the interface between the HDCM and the HDM is a custom discrete wire interface for controlling motion and sensing operations. For example, the control electronics for the purge port tool located in the operating gallery is defined as an HDCM, and interfaces through an umbilical cable to the purge port tool inside the closure cell, which is defined as an HDM. Figure 14 illustrates this relationship. Another example is shown in Fig. 15. The HDCM depicted in the illustration might be the control electronics for the weld tray located in the operating gallery. A weld tray located on the weld carriage would be defined as an HDM.

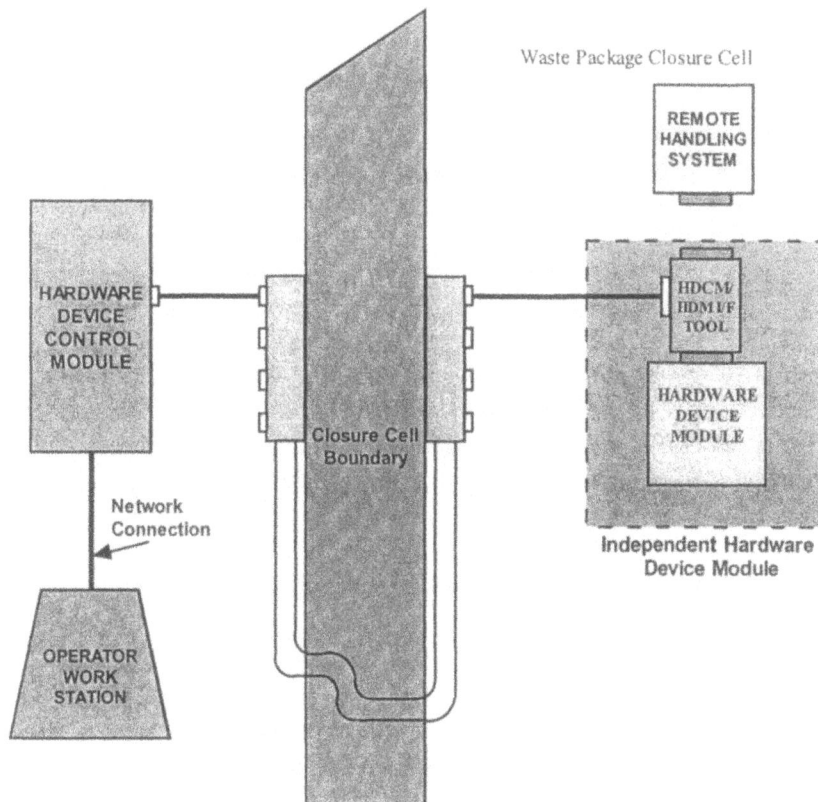

Fig. 14. HDCM/independent module HDM interface (INEEL 2004e).

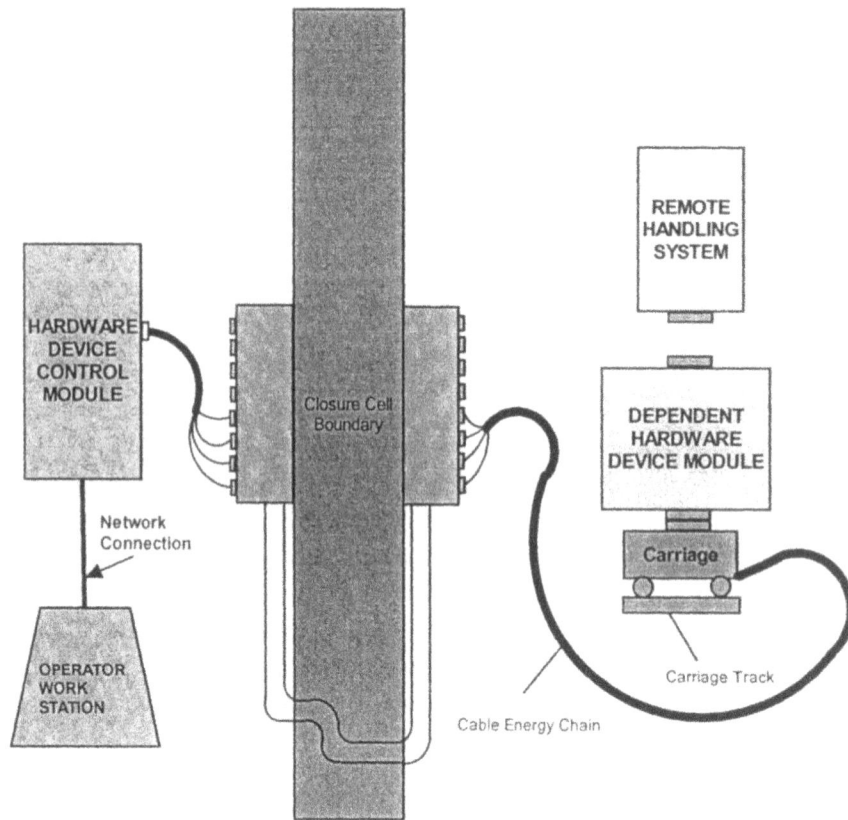

Fig. 15. HDCM/dependent module HDM interface (INEEL 2004e).

The HDMs are defined as either independent modules or dependent modules. An independent module is designed to operate from an umbilical cord interface, and is independent of (and does not derive any control or power from) the weld carriage. A dependent module derives its power and control from the weld carriage located around the closure cell process opening. Both types of modules are placed by the remote handling system (RHS), and both are controlled by an HDCM. The HDCMs are typically located outside the hot cell environment. They may be packaged in NEMA-rated wall-mount enclosures, but more typically are housed in standard 19-in (48-cm) equipment racks. They interface to the in-cell HDMs through standard connector enclosures located on the inside and outside closure cell walls. The block diagram in Fig. 14 depicts the interface between an HDCM and an independent module HDM. All systems and workstations are networked together using Ethernet/TCP/IP. The block diagram in Fig. 15 shows the HDCM/HDM interface for a dependent module. Both independent and dependent HDMs are picked and placed by the RHS, and both have cabling that eventually goes through the standard connector enclosures mounted on the walls of the closure cell. The connection between the HDCM and the OWs is Ethernet (TCP/IP). The wiring through the standard connector enclosures is discrete wiring specific to the HDM requirements.

Wiring from the in-cell connector enclosure connectors to the outer-cell connector enclosure connectors is generic cabling, consisting of twisted-pair, shielded cables. This means that whichever connector is used to pass signals from inside the cell is not critical. It is requisite that correlating connectors on both the inside and outside boxes be specified. Some connectors used on the interface boxes are designed to be remotely connected/disconnected. Umbilical cables used on independent cables may be changed remotely in the event of a failure. Because of the size and weight of the cable management system, cables inside cable management systems associated with dependent modules will be changed-out only through manned entry.

The isometric drawings in Figs. 16 and 17 show the proposed enclosure locations in the operating gallery and the closure cell.

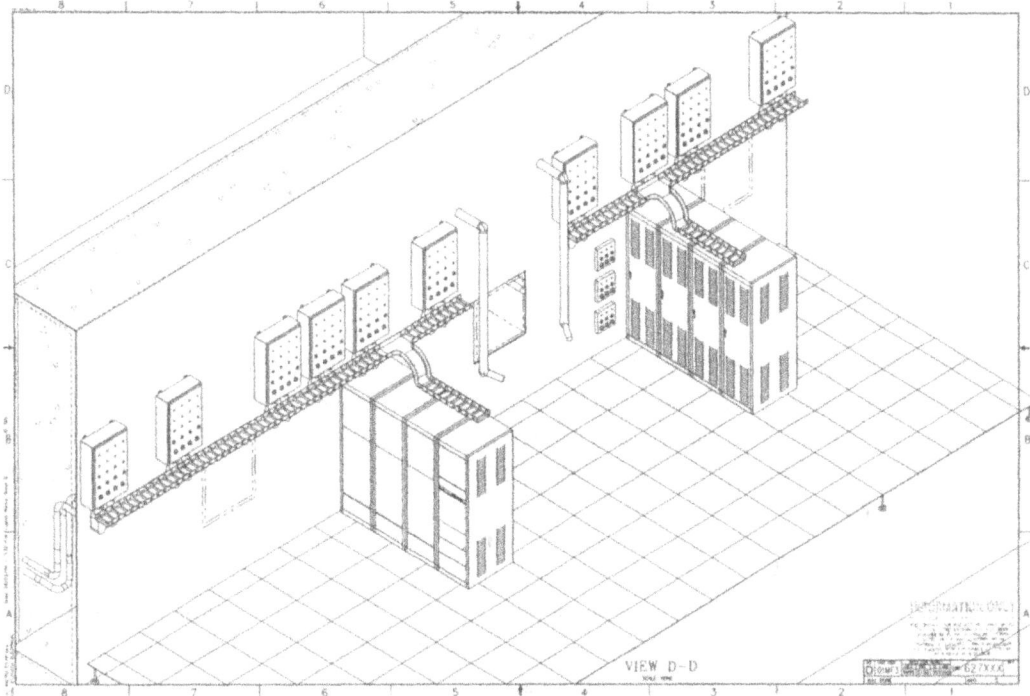

Fig. 16. Operating gallery HDCM/connnector enclosures (INEEL 2004e).

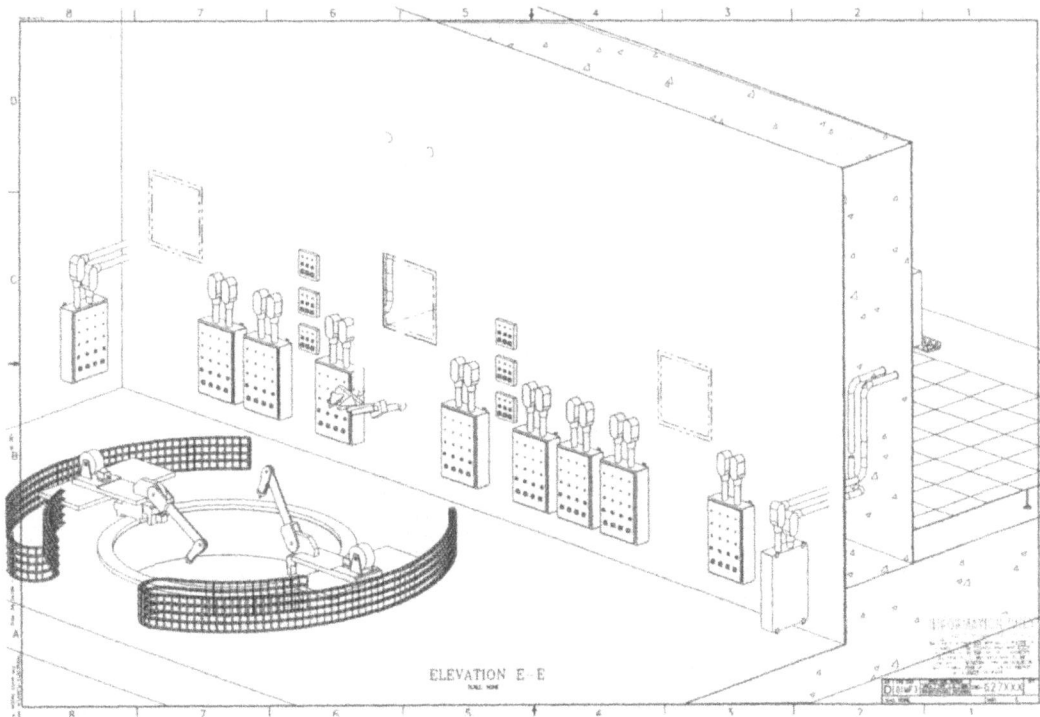

Fig. 17. Closure cell connector enclosures (INEEL 2004e).

Operator workstations

The supervisory control system physically consists of several identical OWs, each dedicated to specific closure cell functions. The closure cell OWs are located in the closure operating gallery, whereas glovebox OWs are located in the closure cell support area. The closure cell OWs have three functions, as follows:

1. **Weld Control:** The welding operator directs the welding operation
2. **Weld Inspection:** The welding inspector certifies that the weld is acceptable
3. **Material Handling:** Operators control equipment in the closure cell, such as the RHS, glovebox doors, etc.

Figure 18 shows an artist's conception of an OW.

The OWs interface through Ethernet (using the TCP/IP protocol) to HDCMs that control in-cell functions. For example, each welding robot is considered to be a hardware device module, and the control electronics necessary to drive the robot is considered to be an HDCM. In this case, the HDCM also directly interfaces to welding processes, which require near-real-time sensing and controls. The HDCM communicates with the OW to provide operator interaction for the welding, and weld inspection activities performed by the HDCM through an Ethernet (TCP/IP) network.

A human-machine interface (HMI) will be developed for each HDCM, and will run on the designated OW. The HMI allows a trained operator to perform the required closure functions with the HDCM, and to perform maintenance and troubleshooting operations.

Camera system

Cameras will be mounted in tubular wall inserts located in the closure cell walls. Video and control cabling will be routed outside the closure cell to the operating gallery video distribution system. Each OW has access to the video signals collected in each operating cell. Operators will have the capability of selecting and displaying up to four of the camera views at their workstations. Machine vision camera views will be routed through dedicated computer systems to provide digital images at the OW. Machine vision systems are used for precise measurement and position determinations.

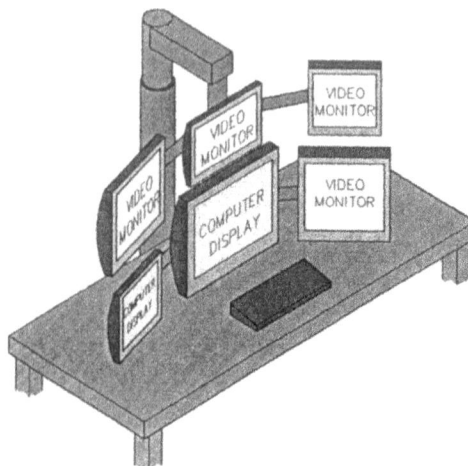

Fig. 18. Layout of a typical operator workstation (INEEL 2004e).

SUPPORT AREA/GLOVEBOX POWER AND CONTROLS INTERFACE

Weld tray control interface stations

Inside each closure cell glovebox, two maintenance areas are designated for repair or refurbishment of the weld trays. When the weld tray is delivered to this area or station by the glovebox RHS, it is placed on a tool plate identical to that located on the closure cell weld carriage. Full functionality of the weld tray is provided through the tool plate connection. This means that a specific area within the support area must be designated for the weld tray control equipment for the weld tray. Space must also be allocated for a welding power supply in the area.

A weld tray tool plate will be located in both maintenance stations. To operate two weld trays simultaneously (one at each station) would require two complete control systems and welding power supplies. If there is no requirement for simultaneous operation, it may be possible to use one control system for both maintenance stations. Cabling would be wired to both quick-change tool plates, but a multiplexing method would allow operation of only one station at a time. The OW local to the glovebox area would be used to control the maintenance station operations.

Independent module control interface stations

An independent module control station is located inside each closure cell glovebox on the side opposite the weld tray control station. This control station is dedicated to maintenance and testing of the independent modules. The RHS is used to transport the independent module from the transfer cart to the glovebox maintenance station. The station mounting structure is identical to the structure on which the module typically mounts in the closure cell. This is a lid-lifting ring in the case of the purge port tool, and it is coupled to an HDCM/HDM interface tool (HHIT), which is identical to that used in the closure cell.

Material-tracking interface station

Several material-tracking interface stations will be used in the closure cell area. Each will interface to the control and data management system (C&DMS) by way of an Ethernet (TCP/IP) interface. The WP traveler documentation will be updated as WP components and quality-related items enter the closure cell support area and are delivered to the gloveboxes.

Safety system interface

The SCS interfaces to the Safety System by way of an Ethernet link using TCP/IP protocol. The support area emergency stop switches; gloveport; glovebox RHS; air lock door; and shield door interlocks are all components of the safety system. The system is designed to provide safe equipment shutdown if an abnormal event occurs. The safety system is an independent system with self-monitoring and diagnostic capabilities.

Operator workstation interfaces

Each operator workstation in the WP support area will communicate through Ethernet (TCP/IP) with the C&DMS. Power will be provided through a local area power outlet to each OW.

OPERATIONS GALLERY/SUPPORT AREA CONTROLS INTERFACE

Camera video link

Several cameras will be mounted in the support area. The link will provide a noise-free signal between the cameras located in the support area and the camera distribution system located in the operating gallery, which will allow operations personnel in the operating gallery to observe operations in the support area.

Voice communications link

An optical fiber communications link will be established between the operating gallery and the support area. This link will interface to wireless communication headsets so that operations personnel can be in constant communications with the supervisory personnel, and with other operators.

Ethernet (TCP/IP) network interface

Glovebox operator workstations will interface with those located in the operating gallery through an Ethernet (TCP/IP) interface. This may be translated through a fiber optic connection, depending on routing requirements.

CONTROL ELECTRONICS EQUIPMENT LOCATIONS

The closure cell HDCMs will be located in the operating gallery, as depicted in Fig. 16. A raised floor will be installed, which will simplify the cabling necessary between the HDCM and operator workstations. Cable lengths from connector interface boxes, bringing signals from inside the cell to electronics located in the operating gallery, must be kept to a minimum. The HDCMs are mounted in standard 19-in (48-cm) equipment racks located on both sides of the WP observation windows.

CONTROL SOFTWARE ARCHITECTURE

The closure cell software control is a collection of software modules. Modules can be added to, or subtracted from, the system without impacting existing modules. A detailed description of each software module is included in the respective software module technical requirements documentation. Each software module will incorporate the following:

1. **Status Reporting:** This includes error code generation; ready/not ready indications; and process status. The format of this information will depend on the complexity associated with the module and the processes being performed. For example, the status reporting for the WP ID end-effector module will be simple ready-to-read and data-collected status. Status for the welding module will include end-effector position; carriage location; robot positions; and other critical items.

2. **Permissives:** These refer to the software controls executed by the supervisory control software to ensure that there are no conflicts and that equipment is operated according to the process flow. The permission for a module to begin operation will be supplied only when all system conditions are correct for

operation of that module, and when the process flow agrees that it is time to execute that module. Hardware and software interlocks, as well as priority level, are considered when permissions to operate are issued.

3. **Data Communications:** These involve transfer of data between software modules and the supervisory control software. The volume of data will vary significantly, depending on the module and the functions it performs.

Access control

The Supervisory Control System (SCS) software will provide security capabilities to limit access to the software codes and system operations. Several permission definitions will be built into the system to provide three levels of access, as follows:

1. A permission level to allow only trained operators to operate the system
2. A higher permission level to allow only trained supervisors to modify the parameters of the system for certain operations
3. An even higher permission level to allow only the systems administrator to access the system security functions

Access control will be maintained using the Windows-based password system. Administrative controls will be required to ensure that passwords are not shared or compromised by operations personnel.

Communications protocol

The SCS provides monitoring and supervisory control functions for all systems within the closure cell. It will interface directly to the OW and to the DCMIS through Ethernet connections. The SCS will provide initial permission for operations to begin in the closure cell. Weld inspection information will be received from the OW, and directed to the appropriate database located either in the SCS or DCMIS.

DCMIS interface

The SCS and the DCMIS will communicate the following information:

1. Notification from the DCMIS to the SCS that a WP is in position at a closure cell process opening
2. Notification from the SCS to the DCMIS that a WP is ready to be moved from a closure cell process opening
3. Status information from the SCS to the DCMIS regarding each WP closure cell
4. The WP closure cell archive data from the SCS to the DCMIS
5. Download of the WP electronic traveler information from the DCMIS to the SCS
6. Validation of WP identity and associated components
7. Validation of operator access to control system functions
8. Upload of updated WP electronic traveler information from the SCS to the DCMIS

CONTROL SOFTWARE COMMUNICATIONS PROTOCOL TO THE HDCMs

The control software communications protocol to the HDCMs has not yet been detailed. The basic outline is that the HDCM will be directed and/or given permission from the SCS to perform a function. As indicated above, by monitoring status, the SCS will know the required configuration and status of equipment sharing the same physical envelope as the specific HDCM being directed, thus precluding conflict or collision. An HDCM is a stand-alone module that will perform a specific function, as directed, and report to the SCS when that function is complete. The Software Module Technical Requirements form is used to describe the details of each software module used in the C&DMS. The form provides the means by which protocol, requirements, interfaces, process flow, and software description are all located in one place for each software module. Each module is assigned a unique number, which is descriptive of the system in which it is used. The numbering format is described on the form.

BOUNDARIES AND INTERFACES

Each station or module communicates through an Ethernet (TCP/IP) interface to a SCS. The software protocol to allow permissive control has not been defined. A portion of the C&DMS will manage data pertaining to the processes in the WP closure cell. The interfaces will include, but are not limited to:

1. Welding and control modules at each WP station
2. Weld inspection modules at each WP station
3. Material tracking systems
4. DCMIS
5. Weld inspection/monitoring consoles at each WP station
6. Weld control consoles at each WP station
7. Waste inventories
8. Materials/supplies inventories
9. Voice communications

PHYSICAL LOCATION, LAYOUT, AND PRINCIPLES OF OPERATION

The C&DMS will be located in the operating gallery on the operating floor. There will be five distinct control consoles at each closure cell location, i.e., two weld inspection/monitoring consoles; two weld control consoles; and the material-handling console. A supervisory control station will also be located in the closure cell operating gallery. The control system architecture is a modular design that can be expanded easily. Standard software protocol will be used for communication among the various modules, with an Ethernet switch providing easy access to the system from any subsystem.

SYSTEM RELIABILITY AND CONTROL FEATURES

Reliability features include a Windows operating (or equivalent) environment; information backup; and data archiving. Each OW will have identical hardware components, providing redundancy, consistency, and flexibility. The OW and supervisory control station will have uninterruptible power supplies to allow controlled shutdown of equipment in the event of power failures. Those items requiring an uninterruptible power supply will be identified during design. Each HDCM operates as an independent control system, with the operator workstation serving as an operator interface. Each is given permission to begin operation from the supervisory control station. Closure operations can be monitored from any SCS console, with appropriate authorization.

OPERATIONS

The SCS will be operated in the closure operating gallery and, therefore, will not be exposed to extreme temperatures or a hazardous environment. It is anticipated that under normal operation conditions, the SCS will be manned with at least one person to oversee operations at all weld stations. The person will have authority to authorize operations within the WPCA. Operation of the SCS will require a level of training that will be verified by the SCS security system before the operator is allowed access. At each closure cell, two weld control consoles will be manned by certified welding operators, and two weld inspection/monitoring consoles will be manned by certified weld inspectors. The SCS will grant permission for a welding operation to commence. Once the welding operation is started, the weld control console operator will have control of all operations relative to his or her weld station, including camera selection and video display. The weld inspector and the weld inspection/monitoring console will work in concert with the weld control console operator to ensure the welds meet the quality level required. When the welding and inspection operation is completed, a weld station status report will be sent to the supervisory control station. The supervisory control station will also be capable of requesting status from any weld station at any time. The system can be shut down in two ways:

1. Through a normal shutdown procedure that will ensure that all equipment is in a safe mode
2. Through an emergency shutdown procedure that immediately removes all positioning power and welding power within a closure cell

Processor power will remain to allow troubleshooting and recovery capabilities. These emergency stop switches will be located at the weld control and weld inspection consoles.

An independent safety system will maintain all interlocks and emergency shutdown capabilities, and maintain a network connection to the SCS. The safety system will be a triple voting system, certified by the appropriate evaluation laboratories (Underwriters Laboratories, Canadian Standard Association, etc.). If an interlock fails, or an emergency stop occurs, the safety system will prevent operation of the equipment (safe standdown and lock) and report the occurrence to the SCS. Inherent within the design of the safety system is constant monitoring and verification of the interlock/emergency switch circuits.

Administrative controls will consist of, as a minimum, password authorization based on training records and supervisory approval. Password authorization levels will permit some to operate the system, while others will be allowed to perform maintenance functions and troubleshooting. Other authorization levels may only allow monitoring of operations.

TESTING AND MAINTENANCE

Each maintenance glovebox associated with the closure cells will have limited module control capabilities that will allow a qualified operator to operate weld system modules while the module is located in the glovebox. This provides a method of testing the module after maintenance, or to power the module as part of required resupply operations. Either the weld control or weld inspection consoles will be capable of providing a maintenance mode of operation at each closure cell. This mode will provide an authorized operator with the capability to move axes of motion; review sensor raw data; and perform other functions relating to troubleshooting. This mode of operation will also allow access to parameter database information relative to the control systems.

DESIGN REQUIREMENTS

As presented above, the C&DMS consists of three levels of control and data management functions. The architecture will be modular, and will use standard commercially-available software to interface with the various hardware and software components in the C&DMS. All three control levels will require custom software, written to industry standard protocol. Where possible, industry standard hardware and software will be used to maximize both flexibility and supportability, while increasing the likelihood of component availability over the life of the project. Table 24 lists the detailed design requirements together with their corresponding performance measures, verification methods, and bases.

Table 24. Control and Data Management System Design Requirements based on the Technical Requirements from TFR-282 (INEEL 2004a).

Requirement Number	Requirement	Performance Measure	Basis	Verification Method
CD-1	The C&DMS shall control all equipment to avoid conflicts within the closure cell.	Control system controls the event state of all operations and communicates with all equipment within the closure cell.	TR-3.6.7-1	Demonstration and test
CD-2	The C&DMS shall provide a means to control the welding and inspection equipment.	At least two stations for each cell to allow human oversight and control of closure cell operations.	TR-3.6.7-1	Demonstration and test
CD-3	The C&DMS shall provide a means to control the material handling equipment.	Operate workstation to allow human oversight and control of closure cell material handling.	TR-3.6.7-1	Demonstration and test
CD-4	The C&DMS shall provide a means to control the visual oversight of closure cell operations by human operators.	View area/volume and (multiple) views at OW.	TR-3.6.2-1, TR-3.6.7-1	Demonstration
CD-5	The C&DMS shall provide a means to acquire, collect, and disseminate all pertinent WP and WP weld-related data.	Collect and store, as a minimum, weld inspection, repair, evacuating, and purging data; TBD.	TR-3.1-5	Observation
CD-6	The C&DMS shall provide data input/output with DCMIS.	A facility processor or simulator will transfer selected data to/from the C&DMS	TR-3.1-5 TR-3.6.2-2	Demonstration
CD-7	The C&DMS shall collect and archive C&DMS process data.	Storage and readout of process data.	TR-3.1-5	Demonstration
CD-8	An Ethernet-linked (TCP/IP) programmable controller will monitor and control miscellaneous functions within the closure cell.	Miscellaneous functions such as tool racks and storage drawers will interface to a PLC or other hardware device to perform specific functions.	TR-3.6.1-12	Test and Demonstration

Table 24. (cont'd).

Requirement Number	Requirement	Performance Measure	Basis	Verification Method
CD-9	The supervisory control system will be able to communicate with other subsystems as needed, using Ethernet (TCP/IP) at 100-Gbit speed.	OW will interface to HDCMs thru Ethernet using TCP/IP protocol	TR-3.6.2-2, TR-3.6.2-3	Demonstration.
CD-10	The supervisory control system will have a means of long-term storage of data. The data storage will be in conjunction with data taken from other subsystems, using a standard storage method (TBD).	Data will be stored on a media yet to be determined.	TR-3.1-5	Demonstration of storage media
CD-11	The C&DMS shall identify and match the lids, caps, and spread rings to the WP.	The Material tracking system will be used to read and verify bumpy barcodes on each of the components listed for a specific WP.	F&OR 1.1.2.3-1	Demonstration of the Material Tracking Subsystem
CD-12	The C&DMS shall access the WP data package.	The C&DMS will collect from a simulator or the DCMIS a copy of the WP data package, add or modify information to the package, and return the package to the DCMIS or simulator.	TR-3.1.5	Demonstration
CD-13	The C&DMS shall provide a means to control the visual oversight of weld cell operations by human operators.	An OW will be used to perform a WP closure or a simulation thereof.	Section 3.2	Demonstration.
CD-14	To the extent possible, electronic equipment shall be located outside of the weld cell.	Operation of the closure equipment.	TR-3.6.7-4	Observation of equipment placement.
CD-15	Windows client/server distributed control software.	Use industry standard control/communications networking software to implement distributed control.	Best engineering practice.	Demonstration
CD-16	The C&DMS shall supervise and control all OWs	The control system controls the event state of all operations and communicates with all equipment within the Weld Cell and gives permission to begin closure process.	Section 3.2	Demonstration.
CD-17	OW (weld control) shall interact with welding HDCMs.	The OW provides the HMI for the welding operations.	Design criteria.	Demonstration

Table 24. (cont'd).

Requirement Number	Requirement	Performance Measure	Basis	Verification Method
CD-18	OW (weld inspection) shall interact with inspection HDCMs	The OW provides the HMI for the inspection operations.	Best engineering practice.	Demonstration.
CD-19	The C&DMS will interface with the DCMIS via 100-Gbit Ethernet using TCP/IP protocol.	C&DMS will communicate WP-related information to the DCMIS, and vice versa.	TR-3.6.2-2	Demonstration
CD-20	The C&DMS will receive notification that WP is in position	DCMIS will control all movement of the WPs and will notify the C&DMS through the Ethernet link that the WP is in position and ready to be welded.	Design. criteria.	Demonstration
CD-21	The C&DMS verifies WP ID per work package information.	The C&DMS will compare the WP ID information from the work package with the information received from the material tracking system.	F&OR 1.1.2.3-1 TR-3.1.5	Demonstration
CD-22	Interfaces to DCMIS for facility alarm notification.	C&DMS will interface with the DCMIS to ensure that facility alarms pertinent to the welding operation are communicated.	Design criteria.	Demonstration.
CD-23	The C&DMS will interface to an independent safety system for closure cell status/alarms.	C&DMS shall be aware of all equipment status, including emergency or interlock conditions.	TR-3.6.7-3.	Demonstration
CD-24	The C&DMS will interface to the glovebox maintenance stations.	Welding equipment is powered and controlled within the maintenance glovebox for troubleshooting, refurbishing, and repair.	Design criteria	Demonstration
CD-25	Support operations with quality materials tracking from refurbishing through WP closure.	Demonstrate materials tracking;	Best engineering practice	Demonstration
CD-26	Waste load-out inventory and control	Inventory list of all items being moved out through the glovebox load-out system.	Design criteria	Demonstration
CD-27	Supervises operations/scheduling	Demonstrate correct closure operation flow scheduling including part(s) identification (lids/split rings), adequate gas supplies, etc.	Best engineering practice	Demonstration

Table 24. (cont'd).

Requirement Number	Requirement	Performance Measure	Basis	Verification Method
CD-28	Manages process data and generates C&DMS reports; archives data	Generate work shift reports, process reports, weld quality verification archival, and TBD.	TR-3.1-5	Demonstration
CD-29	Controls operator authorization/Access	Restrict access to only approved users	Best engineering practice.	Demonstration
CD-30	Supervisory control shall be provided by a server-quality personal computer or network of computers running a recent version of the Microsoft Windows operating system.	Demonstrate that all systems communicate correctly with each other in a client/server environment. The client/server model is implemented using Microsoft's interprocess communication standards. The hardware link will be Ethernet using TCP/IP protocol.	Best engineering practice. Design criteria. Per discussions with BSC at INEEL November 18-19, 2002.	Demonstration
CD-31	Each control module should be connected to the WP closure cell control level network with a standard networking protocol.	Systems communicate in a client/server environment. The client/server model is implemented using Microsoft's interprocess communication standards. The hardware link will be Ethernet using TCP/IP protocol.	TR-3.6.2-3	Demonstration
CD-32	The SCS, in a diagnostics mode of operation, will allow WP closure cell control stations to exercise low-level motion and I/O functions via the Ethernet communication link.	Include Maintenance mode of operation.	Design criteria.	Demonstration
CD-33	Downloading and modification of hardware device control module software shall be possible using a network connection to the supervisory level network.	Provide software hooks and HMI to download HDCM software when necessary.	TR-3.6.2-2	Demonstration
CD-34	Each operation within the closure cell will be monitored and controlled via operator friendly computer display screens.	HMI will be intuitive and easy to use without extensive training.	Design criteria	Demonstration.

Table 24. (cont'd).

Requirement Number	Requirement	Performance Measure	Basis	Verification Method
CD-35	The HMI computers must be capable of displaying operation screens used during normal operation of the closure cell and maintenance/diagnostic screens used for low-level motion and I/O manipulations.	Pull-down menus accessed through password authorization allowing low-level control screens for maintenance and troubleshooting shall be instituted.	Design criteria	Demonstration.
CD-36	C&DMS software modules maintained and controlled under a software quality assurance plan.	Software quality plan PLN-1626 has been instituted.	TR-3.6.8-1	Observation
CD-37	All C&DMS equipment will be designed to withstand the effects of Yucca Mountain natural phenomena in accordance with ICC 2000	TBD	TR-3.6.9-2	Observation
CD-38	C&DMS control equipment will be designed and installed in accordance with NFPA 70 National Electric Code	Adherence of all system/subsystem fabrication and installations to code	TR-3.6.10-1 TR-3.6.11-5 TR-3.6.13-2	Observation/ inspection
CD-39	All in-cell C&DMS components will either be radiation tolerant (hardened), shielded or easily replaceable (when not critical to process operation)	All in-cell interface connectors, cables, and switches will be radiation tolerant; cameras will be designed for operation in radiation areas.	TR-3.6.11-1	Observation/ inspection.
CD-40	In-cell C&DMS components will accommodate decontamination processes to the extent possible.	Connector enclosures and connectors will be sealed and constructed of stainless steel; cable trays will be covered to extent possible and sealed where possible.	TR-3.6.11-2	Observation

The C&DMS interacts with the DCMIS by way of a data communications network (Ethernet using TCP/IP protocol). Operator interfaces are required to oversee and verify operations within the closure cell. The facility will provide power and other utilities required for operation of the C&DMS, as follows:

1. The C&DMS interfaces with the DCMIS by way of an Ethernet link, as defined in the previous requirements. Otherwise, the C&DMS is a stand-alone system.
2. Electrical power outlets for equipment.

Interfaces to facility utilities have not yet been defined. It is anticipated that the following utilities will be required:

1. 120Vac electrical power for the control system equipment
2. Vacuum pumping system
3. Inert gas supply system
4. Demineralized water
5. Instrument air
6. Isolated instrument ground
7. Safety ground

A system qualification test will be performed before onsite system assembly and systems operations (SO) testing. The qualification test will include as much of the system equipment as reasonably possible. This test will be used for design verification of the system control and data recording requirements. An SO test will be performed on the control and data acquisition parts of the system before delivery and customer turnover. All or any other parts of the system may be SO tested at this time. The SO test procedures shall be reviewed and approved by Bechtel SAIC Company, LLC (BSC) before test initiation. An integrated test of the complete system will take place at an Idaho site before turnover to BSC. Control hardware will be designed for remote maintenance or removal using master-slave manipulators or gantry-mounted end-effectors.

The control hardware will be located outside areas exposed to significant radiation levels or protected through shielding, distance, or by minimizing exposure time. All C&DMS systems will be password protected, at a minimum. Remote maintenance and repair will help to reduce exposures, following the ALARA (as low as reasonably achievable) principle. An independent safety system will be used to monitor and mitigate hazards due to electrical equipment operation or failures. It will be UL-listed and certified by the appropriate national and international safety organizations.

Control systems shall be designed to meet the requirements of the National Electrical Code (NFPA 2002a) and Electrical Standard for Industrial Machinery (NFPA 2002b), as well as local, city, and state government requirements. Cable and wire used shall be UL-listed. Transformers shall be-UL listed and CE-certified. Power supplies shall be UL-listed and CE-certified. Motion control equipment shall be UL-listed. Category 3 and Category 4 safety components shall be UL-listed and CE-certified.

WELDING PROCESS: CONTROL FUNCTIONS AND ASSOCIATED PERFORMANCE REQUIREMENTS

WASTE PACKAGE WELDING

The following Information is derived from the EDF-5103, WPCS Welding Process: Control Functions and Associated Performance Requirements (INEEL 2004c) document.

Welding lids to the waste package (WP) vessels will be performed using the cold wire, gas tungsten arc-welding process, which is first set up by tack-welding the lids in place. A spread ring assembly will retain the 316 stainless steel inner lid in position for tack-welding. The spread ring assembly will be tack-welded to the inner lid and the 316 stainless steel inner vessel, and then seal-welded to the inner vessel by a two-pass seal weld. A purge port cap will be placed over the inner vessel purge port, tack-welded, and seal-welded in the same manner as the inner lid-to-spread ring-to-inner vessel tack and seal welds. The Alloy 22 middle lid will be tack-welded, and then multi-pass fillet-welded to the Alloy 22 shell. The outer lid will be tack-welded and narrow groove-welded to the shell. Two welding torches will be used to weld the inner, middle, and outer lids. These torches will be placed symmetrically about the respective weld joints during welding to ensure that residual stresses during welding are reasonably balanced across the lids. This is required to prevent movement of the lids during welding. However, this requirement does not apply for welding the purge port cap to the inner lid.

GAS TUNGSTEN ARC-WELDING PROCESS

Automated or robotic welding with the gas tungsten arc-welding process requires several levels of control. First, the welding torch needs to be positioned in the appropriate location and orientation for welding. Welding a weld joint requires the welding torch to be nominally centered with respect to the welding joint and oriented so that the electrode approximately points into the welding joint. The welding electrode should be located close to the welding joint. Shielding gas is then allowed to flow through the welding torch long enough to ensure that the weld region is adequately shielded. The arc is struck using one of several techniques, and the weld pool is established. The welding torch is moved along the weld joint, and filler wire is added, as needed, to fill or reinforce the weld. The arc voltage may be controlled by varying the distance of the welding electrode to the weld pool, and a current regulator in the welding power supply may be used to provide a pulsed or steady electrical current. The process is stopped at the end of the weld. Starting and stopping of the weld may involve sequencing and ramping welding-control parameters up or down, as appropriate.

CONFIGURATION INFORMATION

Welding torches need to be positioned laterally within about ±0.1 mm (±0.004 in.) of the desired position over the weld joint during welding. The welding torch axis must be positioned within ±1 degree of the desired angular position. Positioning could be done directly by robotic mechanisms or by seam-tracking mechanisms located on the welding end-effectors. In the latter case, the robotic mechanisms would need to position the welding end-effectors within about ±1 mm (±0.04 in.) of the desired position. The welding torches need to be positioned such that the electrode tips will be within about ±0.1 mm (±0.004 in.) of the desired distance (in a direction parallel to the electrode centerline axis) from the weld joint or prior weld bead during welding to maintain a proper arc voltage. Such positioning could be done directly by the robotic mechanisms, or by arc voltage control mechanisms located on the welding end-effectors. If the latter, the robotic mechanisms would need to position the welding end-effectors within about ±1 mm (±0.04 in.) of the desired position.

MOTION CONTROL

Placement of an object at any point in a workspace requires a motion control system having three axes of motion. Placement of the object at a given point, with the object having a given orientation, requires three additional axes of motion. Thus, six axes of motion (or degrees of freedom) are required (McKerrow 1991). These are generally represented as three orthogonal translation axes with rotation around each of the three axes, for a total of six degrees of freedom of motion. Placement of the welding end-effector at any point in (or near) a weld joint may be accomplished using three orthogonal translation axes of motion. This placement will result in the welding torch being maintained in a constant orientation. If the torch were originally oriented with the centerline axis of the torch vertical, that orientation would be maintained during translations. During welding of the outer-lid weld, the torch needs to be in a nominally-vertical orientation. However, for the middle lid and the spread-ring welds, the torch needs to be inclined at nominally 45 degrees from vertical in a plane orthogonal to the welding direction. This inclination requires an additional 4th (rotation) axis of motion. During welding of the spread ring ends to each other, the welding torch needs to be oriented such that the top of the torch is tilted away from the direction of welding in a plane parallel to the welding direction. This requires an additional 5th (rotation) axis of motion. Finally, the welding torch may need to be rotated about its centerline axis to ensure that the filler wire enters the weld joint without the wire guide contacting the weld joint. This requires a 6th (rotation) axis of motion about the weld torch centerline axis.

Simple six-axis robot

Welding of the lids to a WP using one welding torch could be accomplished with a relatively simple six-axis robot. This robot could be built using three commercially-available linear translational motion devices moving a custom-built device having three axes of rotational motion to which the weld end-effector would be mounted. Two such robots would be necessary to correctly position two welding torches that are nominally diametrically opposed across the WP lids. Means would be provided to allow the welding torches to overlap the starts and stops of the weld beads, and to stagger the starts and stops of the weld beads, as is good welding practice. In addition, shielding would be necessary to protect the cables used by the various end-effectors from the significant radiation levels above the WPs. However, a different configuration than the one described above was chosen for the robotic welding system. Justification for this choice is given in EDF-4227 (INEEL 2004g).

Equipment configuration

The basic configuration of the welding equipment will be a circular track machine. The circular track will be mounted a few inches above the operating floor of the closure cell, concentric to a large diameter hole in the operating floor, which will allow the WP to be placed in position below the nominal center of the circular track for welding. Two carriages will be placed on the circular track to move the robots around the WP during welding. Control, data, power, and other utility cables (and hoses) will be run from the closure cell control area to these carriages. A cable management system will allow these cables to follow the carriage motion. The carriages will be capable of motions in excess of 180 degrees around the concentric track, which will allow for overlap of weld bead ends. The carriages will be fitted with commercially-developed 6-axis robotic arms that will allow the welding end-effectors to be placed in position on the WP. The robotic arms will have sufficient range of motion to be positioned to allow lids to be placed on the WP without removing the carriages from the concentric track. The arms will have sufficient reach to allow them to be used to weld the WP purge port cap to the inner lid. A set of end-effectors will be fitted onto each carriage for welding, inspection, and weld repair of the WP closure welds. These end-effectors will connect to the robotic arms by means of quick-release tool change connectors, so that they can be stored in a tool tray on the carriage when not in use. The tool tray will connect to the carriage by means of a quick-disconnect tool plate, and will incorporate a second quick disconnect on top to allow the remote handling system (RHS) to move it to a glovebox in the support area for servicing.

ROBOT ARM

Two identical robotic arms will be used for welding the WP. One robot arm will be mounted to each of the two carriages. Commercial welding robots will be used if possible. The robots will have 6- or 7-axis control. The 7th axis will be optional for carriage motion control. Each robot should have sufficient reach to place the end-effectors in any position between the centerline of the inner lid and the outer rim of the upper lid weld joint. Additional details of the robot arms are given in SPC-572 (INEEL 2004h).

WELDING END-EFFECTOR

One welding end-effector will be used for welding both 316 stainless steel and Alloy 22 components. This will require means to ensure that the appropriate filler wire and shielding gas are used for each weld, and that it is not possible to weld 316 stainless steel using Alloy 22 filler wire and shielding gas, and vice versa. The welding end-effector will incorporate means for quantitative visual inspection of the weld passes, as required. These inspections will be performed using a seam-tracking sensor capable of making horizontal and vertical measurements of both the weld joint and weld surface profiles, and will be mounted ahead of the welding torch. The welding end-effector will incorporate a weld vision camera that will have suitable arc light attenuation capabilities to obtain video images of the weld pool during welding. The camera will be mounted ahead of the welding torch. Each welding end-effector will also incorporate a video camera for video imaging of the weld bead behind the welding torch. The welding end-effector will incorporate a remotely-adjustable filler wire guide to ensure that the filler wire enters the weld pool in such a manner as to facilitate making a good weld. The welding end-effector will incorporate a thermocouple or other temperature sensor capable of measuring the temperature of the base metal before each weld pass. The welding end-effector will incorporate a mechanism to provide lateral motion of the weld torch normal to the welding direction. This mechanism will be used for seam-tracking and weld-torch oscillation during welding. The seam-tracking sensor discussed above will be used to detect a seam tracking error. The welding end-effector will incorporate a mechanism to enable vertical motion of the weld torch parallel to the axis of the torch. This mechanism will be used for automatic voltage control during welding and arc touch starting.

MOTION TRAJECTORIES

The inner lid of the WP is held in place structurally with a spread ring, but the spread ring must be sealed to both the inner lid and the inner vessel. First, a set of tack welds will be made between the spread ring and the inner lid, and between the spread ring and the inner vessel. Then, two single-pass seal welds will be made between the spread ring and the inner vessel (one on each side, for 180 degrees of circumference, plus some overlap at each end); two single-pass seal welds will be made between the spread ring and the inner lid; and two single-pass seal welds will be made between the two ends of the spread ring. The purge port cap will be welded to the inner lid with tack welds made between the purge port cap and the inner lid. Then, two single-pass seal welds will be made between the purge port cap and the inner lid.

The middle lid will be fillet-welded to the Alloy 22 shell. Several tack welds will be made between the middle lid and the Alloy 22 shell. Then, several weld passes will be made between the middle lid and the Alloy 22 shell on each half of the WP, with overlap of the weld pass ends. The outer lid will be narrow-groove welded to the Alloy 22 shell. A set of tack welds will be made between the outer lid and the Alloy 22 shell. Then, the outer lid will be welded to the Alloy 22 shell with multiple weld passes between the outer lid and the Alloy 22 shell on each half of the WP, with overlap of the weld pass ends. Weld starts and stops need to be over-lapped to prevent them from stacking up over each other.

In addition, robotic trajectories typically involve rapid motion from a home position to a point near the weld start point; slower movement to the start point; controlled tracking motion to the weld end-point; slow retraction to a point near the stop point; and then rapid motion back to the home position. The result of the above operations is that a very large set of unique 3-dimensional points must be identified in the robot workspace over the WPs solely for welding. These points define various start and end-points for trajecto-ries of weld torch motion. Additional details of the welding process are given in EDF-4278 (INEEL 2004f).

DESIGN DESCRIPTION

Commercial off-the-shelf (COTS) control system components will be used to monitor and control the various functions of the welding system, when available.

User interface

A human/machine interface (HMI) will allow human operators to operate and interact with the welding high-level motion trajectory planning and process controls system as an integral part of the outer-loop control structure. The HMI will incorporate a graphical user interface (GUI) displayed on one or more computer screens—with additional control devices, such as a joystick; mouse; track ball; touch screen; and inter-operator voice communications—within the closure cell operator workstations. Manual operator trajectory and process-control offsets will be implemented through the HMI and GUI interfaces as part of the outer-loop trajectory guidance algorithms. These operator offsets are transmitted as set-point or baseline trajectory offsets onto lower level set-point and tracking controllers operated within the welding control subsystems.

Control computer architecture

The welding system control architecture will be segmented into two primary levels, as follows:

1. Low-level set-point and trajectory-based trackers, implemented within subsystem control components or within the weld system control computers
2. Higher-level motion trajectory planning and process controls in which the welding operators will be an integral part of the control loop design, implemented within the closure cell operator station computer system

Subsystem controllers will be utilized to the greatest intent practical, while higher-level control functions will be provided in the operator interface and supervisory control system (SCS) functions. At present, the welding computer subsystems consist of rack-mountable PC-type computers with direct interfacing to welding subsystem components via Ethernet; RS-232; video; analog; etc., as appropriate. This approach segments the low-level set-point and trajectory—following control algorithms away from higher GUI-based motion and process-control algorithms executed within the closure cell operator workstations. It is within the higher-level GUI-based control algorithms that the operator feedback loops and direct operator override control will be implemented. Application software will be developed in various programming environments and languages using LabView; C; C++; Java; or some combination thereof. Ethernet and TCP/IP, or similar communication protocols, will be used for inter-system level communications.

Carriages

The carriage motor will be a commercial grade unit, with appropriate motor controllers. The motor controllers will have appropriate computer interface standards, such as Ethernet or RS-232/RS-422/RS-485, for connection to either the robot controller or an independent PC-type computer control. Carriage travel speeds will be between 0 and 200 inches per minute (0 to 5 m/min). Accelerations for the lid welds will be based on welding speeds between 0 and 10 inches per minute (0 to 25 cm/min), measured at the welding torch tip. Accelerations for the purge port welds will be based on welding speeds between 0 and 5 inches per minute (0 to 12.5 cm/min), measured at the welding torch tip. The tool tray will be mounted to the carriage by means of a remotely-operable quick-disconnect tool plate. Most of the electrical conductors and utilities to the end-effectors will pass through quick-disconnect plugs on the tool plate. A second remotely-operable quick disconnect will be mounted to the upper part of the tool tray as a means for the RHS to move the tool tray to the glovebox for service and maintenance.

Robotic manipulator arms

The robotic manipulator arm will be a commercial grade radiation hardened unit, which will include a commercial grade positional controller as specified in SPC-572 (INEEL 2004h). The controller will interface with the weld process control computer by way of an appropriate computer interface standard, such as Ethernet or RS-232/RS-422/RS-485.

Tool trays

Each carriage will incorporate a remotely-removable tool tray. The tool tray will be used for storing end-effectors, dressing tools, and electrodes. It will also hold the filler wire feeders; filler wire spools; gas cup/electrode stickout adjustment devices; electrode changer; dressing tool changers; and possibly one or more video cameras.

Welding end-effectors

The welding power supply will be a commercial grade unit incorporating pulsed current controls. The power supply will incorporate an analog or Ethernet input interface, which will allow pulsed current waveform parameters to be specified from an external source, and an independent PC-type computer. The outer-loop set-point control and setup of this unit will be accomplished via the welding procedures and weldstation operator's HMI and GUI applications. Output power will be 5 - 40 V at 20 - 500 A at 100% duty cycle. Pulsed waveform duty cycle ranges from 0 to 100%. Waveform frequencies bandwidth should be from 0 to 20 Hz. The contactor will be externally controlled.

The wire feeders will supply welding wire to the welding process. They will be radiation-hardened, commercial grade units controlled via an analog or digital interface, by either the welding power supply and/or under independent PC-type computer control. The weldstation operator, via the HMI and GUI applications under designated welding procedures, will primarily maintain the outer-loop control for these devices, with one main exception: automatic fill control. During welding operations, where automatic fill control is enabled, the wire feeders will be under direct control of the adaptive fill controller. The diameter of the wire should range from 0.035 to 0.0625 inches (0.9 mm to 1.5 mm). The wire feed speed should range from 25 to 500 inches per minute (0.63 to 12.65 m/min).

The filler wire position control will comprise two parts:

1. The high-level weldstation operator-controlled GUI for setting and maintaining outer-loop set-points
2. The low-level hardware/software regulator module

The weldstation operator will be presented a live image of the weld wire with respect to the weld pool, so that the weld wire position can be maintained in an optimal welding position based on user experience and existing welding procedures. In this case, the weldstation operator acts as the outer-loop controller for wire positioning. Constant monitoring of the weld-wire-to-weld-pool relationship will be needed due to its coupling with the automatic voltage control algorithm (AVC). Longitudinal travel is along the weld axis of 1.0 inch (2.5 cm), measured at the tip of the wire where it enters the weld pool. Lateral travel is across the weld axis of 1.5 inches (3.8 cm), measured at the tip of the wire where it enters the weld pool.

The cross-axis positioner and oscillation controller will consist of a radiation-hardened, commercial grade linear slide unit controlled via an analog or digital interface, by either the robot controller or by independent PC-type computer control. Furthermore, the seam-tracking trajectory planner will provide sensory input to this controller concerning oscillation endpoint and weld centerline feedback. The weldstation operator, via the HMI and GUI applications under designated welding procedures, will primarily maintain offset adjustment controls for this device in conjunction with the tracking trajectory planner. Lateral travel for the weld process across the axis of the weld will be at least ±0.5 inch (±13 mm). Lateral travel across the axis positioner will be at least 4.5 inches (11.4 cm). Oscillation frequencies will be allowed within the range of 0 to 5 Hz. Independent dwell times (position hold times that occur at each extreme travel location within the oscillation path) will be within the range of 0 to 1 seconds.

Weld torch standoff control will be accomplished by a commercial grade linear slide controlled via an analog or digital interface, by either the robot controller or under independent PC-type computer control with a standoff feedback loop containing the automatic voltage controller. The weldstation operator, via the HMI and GUI applications under designated welding procedures, will have override adjustment controls over this device. Travel will be at least 4.9 inches (12.4 cm).

The seam tracker will be commercial grade. Having been given the nominal weld joint geometry from the welding procedure, this unit will return geometric information pertaining to the measured weld joint centerline, as well as overall measured geometric weld joint parameters (e.g., weld joint outline). It will present the information in such a form that weld joint geometry might be evaluated (e.g., width, depth). The unit will provide feedback on the weld geometry prediction, i.e., if the seam-tracking algorithm cannot determine the weld joint geometry during any frame, it will report this fact, not simply estimate a result. This information will be transmitted to the tracking trajectory planner for automated seam-tracking. The seam tracker will have a minimum resolution of 0.016 inches (0.4 mm). Update frequencies will be nominally 30 to 60 Hz.

The conjectured tracking trajectory planner will consist primarily of a software-based outer-loop trajectory controller implemented within the robot controller or weld control computer. The trajectory planner will search for a weld joint location within a predetermined location and area on the WP, coordinated via general WP coordinates provided by the closure cell SCS. The planner searches using the seam-tracking sensor and weld end-effector cameras assisted by the weldstation operator via the GUI system. Once the nominal weld joint location has been determined within the robot's coordinate frame, the robot and carriage baseline trajectories will be generated based on expected WP geometry. Using this set of baseline trajectories, coupled with the seam-tracking sensor package, a simulated welding pass will be preformed via the robot and carriage controllers motion control systems. During this simulated welding pass, an actual baseline joint trajectory will be determined using the measured weld joint locations and geometries from the seam tracker's feedback, in conjunction with the predetermined baseline trajectory being executed. This new baseline trajectory will be updated (if needed) on each successive pass, based on new weld joint measurements; and augmented by welding procedural bead placements, with allowable adjustments by the weldstation operator via the GUI. In the case of the outer lid weld, the currently predicted weld pass joint geometry and location will be passed on to the adaptive fill controller for weld parameter planning and control activities, which cover the first hot pass weld bead through the final fill pass weld bead just below the cap pass weld bead. Furthermore, during the last groove weld pass on the outer lid weld, the baseline trajectory will be frozen and seam tracking will be discontinued, because no consistent automatically-detectable weld joint will be visible after this weld pass has been made.

Adaptive fill control will primarily consist of a software-based controller implemented within the robot controller or weld control computer. The adaptive fill algorithm's primary use will be for outer lid welding. It will plan the weld bead depths and welding parameters for the first hot pass through the final weld pass just before the cap pass, which consists of several passes. Using weld joint trajectory geometry, determined by the trajectory planner's baseline trajectory and measured weld joint geometry, a welding plan will be developed for each pass, which consists of a series of welding parameter sets (e.g., bead pass height, wire feed speed, heat input) along the joint. This welding plan will ensure uniform fill height on a pass-by-pass basis. The welding plan will be generated at uniform intervals along the weld joint, using a simple rule-based procedure that will develop welding parameters through a simple set of calculations and interpolations on a set of validated welding parameters for a predetermined set of weld joint geometries. The procedure will include simple calculations and lookup tables. One calculation that might be included would choose the target wire feed speed for a measured weld joint geometry by estimating the required volume of weld wire needed within that joint for proper bead height via a simple relationship between filler wire diameter, wire feed speed, and travel speed. Additionally, one possible lookup table might relate the travel speed, required fill deposition, and joint geometry to a validated weld heat input. It is assumed that initial travel speed will largely remain constant throughout each weld pass. When executed, the plan will be interpolated along the full length of the weld, ensuring smooth transitions between weld plan changes. Provision for weldstation operator adjustments to welding plan parameters and real-time control will be provided through the operator workstation's GUI.

Two possible implementations exist for an AVC within this design, as follows:

1. To use an AVC within the robot controller, if it exists
2. To implement an AVC within the software

It is assumed that the robot controller lacks an AVC. Therefore, the AVC will primarily consist of a software-based controller implemented within the robot controller or weld control computer. This controller uses the weld torch standoff controller's set point to maintain a nominally constant welding voltage, given changing weld joint parameters and geometries. The basic controller will be a simple tuned proportional integrative derivative (PID) controller, with a slow digital control loop update time (on the order of seconds or tens of seconds), having:

1. a voltage control resolution of ±0.05 V
2. a controlled voltage range within 8 to 20 V
3. updated frequencies ranging between 0 and 20 Hz

Touch start control will primarily consist of a software-based controller implemented within the robot controller or weld control computer. Two primary current control methods are available to the touch start control algorithm, as follows:

1. Touch start hardware level controller embedded within the welding power supply
2. Software-based preplanned current limiting trajectory that can be used to issue a staged approach to the weld supply control unit

The weld touch standoff controller will be used to slowly move the weld torch electrode toward the base metal while monitoring either the voltage or current output of the current-limited welding supply. Once an appropriate change in baseline values for the welding supply has occurred, an arc start feasibility will be assumed. Next, a predetermined liftoff and current modulation procedure will be executed, ensuring adequate stability of the welding arc throughout its duration. Arc initiation will be verified and reported to the weldstation operator GUI, and the supervisory controller will take over.

The base metal temperature sensor will consist of a commercial grade thermocouple-based temperature-sensing instrument interfaced to the weld control computer via a suitable interface (e.g., analog, Ethernet, RS-232). The thermocouple will be attached to the weld end-effector, deployed onto the surface of the WP, and allowed to equilibrate before the WP inner pass temperature measurements are taken. The inner pass temperature measurements will be sent to the weldstation operator's GUI for evaluation to issue a clear-to-weld temperature release, at which point the sensor will be retracted. The thermocouple will be contact type with a 0 to 500°F (-18 to 260°C) full-scale range, and will have an accuracy of ±1.8°F (±1°C) or ±1.35% of temperature sensed, whichever is greater.

WELDING TORCH POSITIONING

The location of the WP in the closure cell is given in TFR-282 (INEEL 2004a), which gives tolerances on the WP location (e.g., X-offset, Y-offset, Z-offset, and tilt).

Purge port cap

The positioning of a welding torch to weld a purge port cap is shown in Fig. 19. Assuming the type of WP is known, and thus its diameter is known, the uncertainty associated with the weld torch position may be calculated. The welding torch position uncertainty ΔP is given by:

$$\Delta P(X, Y, Z) = X\text{-offset}(\pm 2.0 \text{ in}) + Y\text{-offset}(\pm 2.0 \text{ in}) + Z\text{-offset}(\pm 2.0 \text{ in}) + \text{tilt}(\pm 0.5 \text{ in.})(r/R)$$

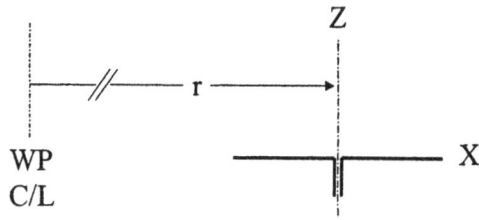

Fig. 19. Sketch of a purge port cap weld joint at radius r from the center of the waste package (INEEL 2004c).

where: X, Y, and Z are Cartesian coordinates with X and Y in the plane of the closure cell floor and Z parallel to the WP centerline; r is purge port cap radius; and R is WP radius bounded by the maximum distances given by X-offset, Y-offset, Z-offset, and tilt. Assuming r/R is a small number, the ΔP values are:

$\Delta P(X) = (\pm \text{offset}) = \pm 2.0$ in. (± 5 cm)
$\Delta P(Y) = (\pm \text{offset}) = \pm 2.0$ in. (± 5 cm)
$\Delta P(Z) = (\pm \text{offset}) = \pm 2.0$ in. (± 5 cm)

Outer lid

The positioning of a welding torch to weld the outer lid joint is shown in Fig. 20. Assuming the type of WP is known and thus its diameter is known, the uncertainty associated with the weld torch position may be calculated. The welding torch position uncertainty ΔP is given by:

$\Delta P = \text{X-offset}(\pm 2.0 \text{ in}) + \text{Y-offset}(\pm 2.0 \text{ in}) + \text{Z-offset}(\pm 2.0 \text{ in}) + \text{tilt}(\pm 0.5 \text{ in.}) \pm w + h - d$

where: X, Y, and Z are Cartesian coordinates with X and Y in the plane of the closure cell floor and Z is parallel to the waste package centerline; R is weld joint radius; w is weld joint width; d is weld joint depth; and h is maximum bead reinforcement height bounded by the maximum distances given by X-offset, Y-offset, Z-offset, and tilt. Assuming w = ± 1.0 in. (to allow adequate width for the weld cover pass), d = -1.0 in.; h = 1/16 in.; and the ΔP values are:

$\Delta P(X) = (\pm \text{offset}) + (\pm \text{joint width}) = (\pm 2.0) + (\pm 1.0) = \pm 3.0$ in. (± 7.5 cm)
$\Delta P(Y) = (\pm \text{offset}) = \pm 2.0$ in. (± 5 cm)
$\Delta P(+Z) = (+\text{offset}) + (\text{reinforcement}) + (+\text{tilt}) = (2.0) + (0.0625) + (0.5) = +2.6$ in. (6.6 cm)
$\Delta P(-Z) = (-\text{offset}) + (\text{joint depth}) + (-\text{tilt}) = (-2.0) + (-1.0) + (-0.5) = -3.5$ in. (-8.9 cm)

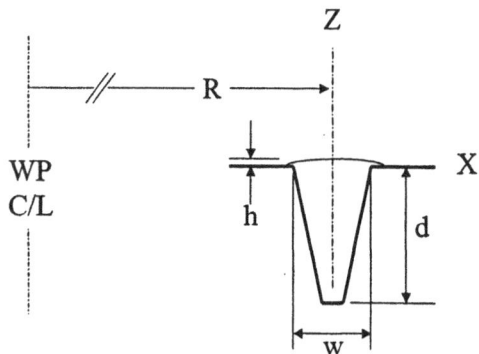

Fig. 20. Sketch of outer lid weld joint at radius R from center of waste package (INEEL 2004c).

111

Middle lid and spread ring

The positioning of a welding torch to weld the middle lid joint or spread ring is shown in Fig. 21. Assuming the type of WP is known, and thus its diameter is known, the uncertainty associated with the weld torch position may be calculated. The welding torch position uncertainty ΔP is given by:

$$\Delta P = \text{X-offset}(\pm 2.0 \text{ in}) + \text{Y-offset}(\pm 2.0 \text{ in}) + \text{Z-offset}(\pm 2.0 \text{ in}) + \text{tiltZ}(\pm 0.5 \text{ in.}) + (\lambda)Z + (\lambda)X.$$

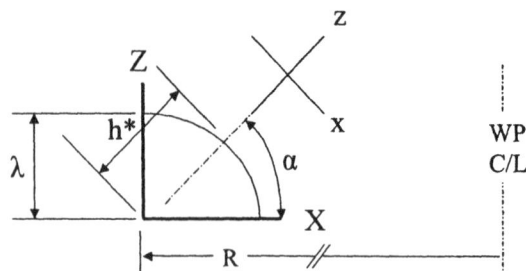

Fig. 21. Sketch of middle lid joint at radius R from center of waste package (INEEL 2004c).

where: X, Y, and Z are Cartesian coordinates with X and Y in the plane of the closure cell floor and Z is parallel to the WP centerline; R is weld joint radius; w is weld joint width; α is torch from horizontal; λ is weld leg length; and h^* is weld throat bounded by the maximum distances given by X-offset, Y-offset, Z offset, and tilt. Assuming λ is 7/8 in, the ΔP values are:

$\Delta P(+X) = (+\text{offset}) + (\text{leg length}) = (+2.0) + (0.875) = 2.9$ in. (7.3 cm)
$\Delta P(-X) = (-\text{offset}) = -2.0$ in. (-5 cm)
$\Delta P(\pm Y) = (\pm\text{offset}) = \pm 2.0$ in. (± 5 cm)
$\Delta P(+Z) = (+\text{offset}) + (+\text{leg length}) + (+\text{tilt}) = (+2.0) + (0.875) + (0.5) = 3.4$ in. (8.6 cm)
$\Delta P(-Z) = (-\text{offset}) + (-\text{tilt}) = (-2.0) + (-0.5) = -2.5$ in. (-6.3 cm)

Welding end-effector slide motion

During actual welding of the middle lid or spread ring, the weld torch will be tilted at about 40 to 45 degrees from the horizontal (see α in Fig. 21). Two linear slides will be incorporated in the welding end-effector to:

1. move the weld torch laterally relative to the weld joint for seam tracking and torch oscillation.
2. move the torch toward or away from the weld bead to control arc voltage.

The corresponding axes of motion are designated x and z in Figs. 21 and 22. The $\Delta P(X, Z)$ values are taken from the above "Middle lid and spread ring" section. The $\Delta P(Y)$ values are not considered because the end-effector slides will not produce torch motion in the Y direction.

For $\Delta P(X) = 2.875$ in., assuming α is 40° (Fig. 22):

$\Delta x = 2.875 \cos (90°-\alpha)$ in. and $\Delta z = 2.875 \sin(90°-\alpha)$ in.
$\Delta x = 2.875 \cos (50°)$ in. and $\Delta z = 2.875 \sin(50°)$ in.
$\Delta x = (2.875)(0.6428)$ in. and $\Delta z = (2.875)(0.7660)$ in.
$\Delta x = 1.8$ in. (4.6 cm) and $\Delta z = 2.2$ in. (5.6 cm)

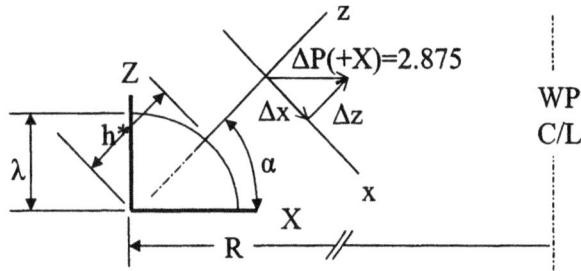

Fig. 22. Sketch showing geometry for Δx and Δz calculations (INEEL 2004c).

For ΔP(X) = -2.0 in., assuming α is 40° (Fig. 23):

Δx = -2.0 cos(90°-α) in., and Δz = -2.0 sin(90°-α) in.
Δx = -2.0 cos(50°) in., and Δz = -2.0 sin(50°) in.
Δx = (-2.0)(0.6428) in., and Δz = (-2.0)(0.7660) in.
Δx = -1.3 in. (-3.3 cm), and Δz = -1.5 in. (-3.8 cm)

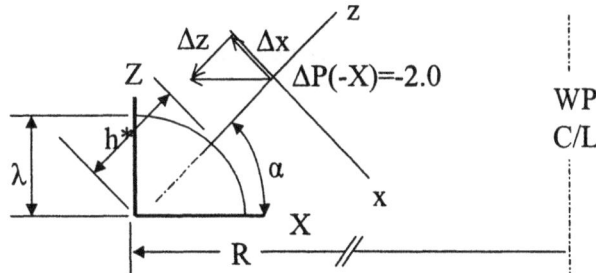

Fig. 23. Sketch showing geometry for Δx and Δz calculations (INEEL 2004c).

For ΔP(Z) = 3.375 in., assuming α is 40° (Fig. 24):

Δx = -3.375 cos(α) in., and Δz = 3.375 sin(α) in.
Δx = -3.375 cos(40°) in., and Δz = 3.375 sin(40°) in.
Δx = (-3.375)(0.7660) in., and Δz = (3.375)(0.6428) in.
Δx = -2.6 in. (-8.6 cm), and Δz = 2.2 in. (5.6 cm)

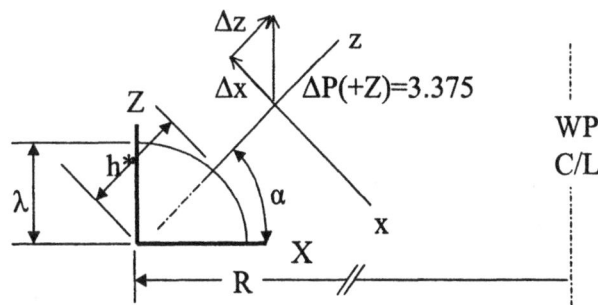

Fig. 24. Sketch showing geometry for Δx and Δz calculations (INEEL 2004c).

113

For $\Delta P(-Z) = -2.5$ in. (down), assuming α is 40° (Fig. 25):

$\Delta x = 2.5 \cos(\alpha)$ in., and $\Delta z = -2.5 \sin(\alpha)$ in.
$\Delta x = 2.0 \cos(40°)$ in., and $\Delta z = -2.0 \sin(40°)$ in.
$\Delta x = (2.0)(0.7660)$ in., and $\Delta z = (-2.0)(0.6428)$ in.
$\Delta x = 1.5$ in. (3.8 cm), and $\Delta z = -1.3$ in. (-3.3 cm)

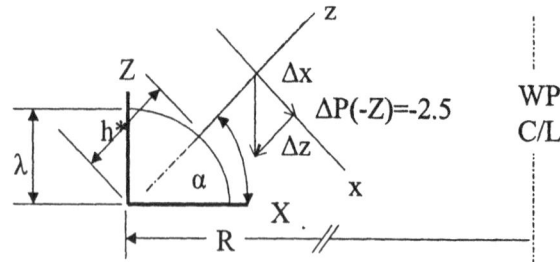

Fig. 25. Sketch showing geometry for Δx and Δz calculations (INEEL 2004c).

Automatic voltage control range of motion

The range of motion to accommodate automatic voltage control for the z-axis is nominally λ plus ¼ in. (0.6 cm), or $\Delta x = h^* + 0.250 = 0.875 + 0.250 = 1.2$ in. (3 cm).

Bounding cases

From the above calculations the maximum position deviations are:

$\Delta P(X) = \pm 3.0$ in. (± 7.5 cm)
$\Delta P(Y) = \pm 2.0$ in. (± 5 cm)
$\Delta P(+Z) = 3.4$ in. (8.6 cm)
$\Delta P(-Z) = -3.5$ in (-8.9 cm)

The minimum range of travel needed for the seam-tracking slide is $\Delta x = 1.848 - (-2.585)$ in. $= 4.5$ in. (11.4 cm). Less automatic voltage control, the maximum range of travel needed for the automatic voltage control slide, is $\Delta z = 2.202$ in. $- (-1.532)$ in. $= 3.8$ in. (9.6 cm). Including automatic voltage control and torch motion for seam tracking, the minimum range of travel needed for the automatic voltage control slide is nominally $\Delta z = 3.734$ in. $+ 1.125$ in. $= 4.9$ in (12.4 cm).

Robotic solutions

The following three approaches are considered for control-welding torch positioning:

1. **Fixed Robot Position:** In this approach, the carriage would move to provide motion of the welding torch along the particular weld joint, and the robot would be prepositioned at a location nominally centered with respect to the weld to be made. The robot would not subsequently move during welding. The seam tracking and automatic voltage control slides on the welding end-effector would provide all other required motion of the torch. It is not intended to use this approach. The seam-tracking sensor has been sole-sourced from Servo-Robot Inc., which recommends that the sensor be mounted to the body of the welding

end-effector—not to the welding torch. The seam-tracking sensor does not have an adequate field of view to detect the weld joints for all possible WP locations in the closure cell when mounted to the body of the end-effector.

2. **Preplanned Robot Position:** In this approach, the carriage would move to provide motion of the welding torch along the particular weld joint, and the robot would be preprogrammed to move in seven degree-of-freedom coordination with the carriage on a circular, elliptical, spline, or other appropriate trajectory chosen or calculated to result in close approximation to the trajectory needed to weld a particular weld joint. The seam tracking and automatic voltage control slides on the welding end-effector would provide correction for deviation from the chosen or calculated torch trajectory, and the trajectory actually needed to make the weld. This is the approach selected for implementation.

3. **Dynamic Robot Position:** In this approach, the carriage would move to provide motion of the welding torch along the particular weld joint, and the robot would be programmed to move in nine degree-of-freedom coordination with the seam tracking and automatic voltage control slides on the welding end-effector and the carriage, to provide correction for torch trajectory needed to make the weld. The seam-tracking sensor would provide error signals during welding for use by both the robot and the two end-effector slides. Due to the complexity of the coordinating motion of the robot, and of the two end-effector slides during welding, it is not intended to use this approach.

General robot trajectories

Graph notation will be used for ease of representation of complex actions and trajectories. An action being taken while a robot is stationary, such as engagement of a quick-release connector, can be represented as a graph, where the shading indicates that some action is taken at a point **a** (Fig. 26). Using graph notation, movement along a trajectory, T, from any point **a** to any point **b** may be represented as in Fig. 27. Figure 28 shows a similar representation in which motion along trajectory T_w occurs from any point **a** to any point **b**, and some action occurs represented by the bold arrow (e.g., welding) in addition to motion along the trajectory. Finally, a double-headed arrow is used to represent rotation about an axis (Fig. 29).

a

Fig. 26. Graph representing action at a point (INEEL 2004c).

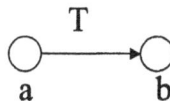

Fig. 27. Graph representing motion from point a to point b (INEEL 2004c).

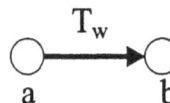

Fig. 28. Graph representing motion from point a to point b, with additional action occurring along the trajectory from point a to point b (INEEL 2004c).

a

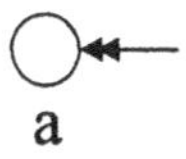

Fig. 29. Graph representing rotation about an axis at some point a (INEEL 2004c).

115

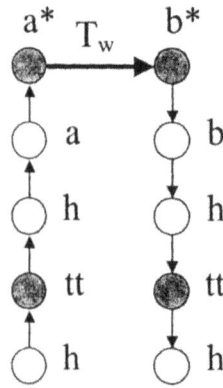

Fig. 30. Graph representing a set of complex motions and actions (INEEL 2004c).

Using these primitives, but not the rotation primitive, a complex procedure is represented in Fig. 30. In this representation, the robot is moved from home position, **h**, to the tool tray location, **tt**, of the welding end-effector. The quick disconnect is engaged between the robot arm and the welding end-effector at **tt**; robot then moves back to the home position, **h**. From there, the robot moves to a standoff position, **a**, over the outer lid weld joint, and then moves slowly, with reference to seam-tracking sensor data, to position **a*** in the outer lid weld root. At **a***, the robot strikes the welding arc and welds along the trajectory T_w. At the end-point of the weld, **b***, the weld arc is extinguished. From **b***, the welding torch raises to standoff position **b**, and then the robot moves to its home position, **h**. Next, the robot moves back to the tool tray position, **tt**, for the welding end-effector, and disengages the quick disconnect at **tt**. Finally, the robot moves back to the home position, **h**.

During closure of a WP, several hundred such procedures will be executed. In order to define these procedures, they will be embedded in external text documents. In these documents, each primitive will be defined by a text string giving critical parameters, such as coordinates in robot space; trajectories in tool coordinates; specific actions to be taken; and associated welding process parameters. This approach will allow separating procedure definitions from robot control programs in a manner that should make both software development and procedure development relatively easy to manage.

CONTROL SYSTEM ARCHITECTURE

There are at least two major options for structuring the control system architecture:

1. **Control Through the Robot Controller:** In this approach, which is expected to be used, as many of the control functions as possible would be done using the robot controller. Most of the control functions will either be supplied with the robot controller software, or can readily be programmed as control macros in the robot controller software. Associated robot requirements are given in SPC-572 (INEEL 2004h).

2. **Control External to the Robot Controller:** In this approach, many of the control functions would be done using a computer and various control hardware external to the robot controller. This approach will be used only if it becomes apparent that control through the robot controller is not practical.

References

AGS (American Glovebox Society). Guideline for gloveboxes. 2nd edition, AGS-G001. Santa Rosa, CA: American Glovebox Society; 1998.

Allen, S. Software quality assurance plan for the waste package closure system Yucca Mountain Project, PLN-1490, 05/06/04. Idaho Falls, ID: Idaho National Engineering and Environmental Laboratory; (May 6) 2004.

ANSI (American National Standards Institute). Safety in welding, cutting, and allied processes, ANSI Z49.1. New York, NY: American National Standards Institute; 1973.

ASTM (American Society for Testing and Materials). Guide for radiation shielding window components used in hot cell, ASTM WK602. Draft under development. West Conshohoken, PA: ASTM International; 2003.

ASME (American Society of Mechanical Engineers). Boiler and pressure vessel code. New York, NY: ASME; 2001.

ASME (The American Society of Mechanical Engineers). Manual for peer review. Washington, DC: ASME; 2003.

ASME/RSI (The American Society of Mechanical Engineers/Institute for Regulatory Science). Assessment of technologies supported by the U.S. Department of Energy; results of the peer review for fiscal year 1997, CRTD Vol. 47. New York: ASME; 1997.

ASME/RSI (The American Society of Mechanical Engineers/Institute for Regulatory Science). Assessment of technologies supported by the U.S. Department of Energy; results of the peer review for fiscal year 1998, CRTD Vol. 50. New York: ASME; 1998.

ASME/RSI (The American Society of Mechanical Engineers/Institute for Regulatory Science). Assessment of technologies supported by the U.S. Department of Energy; results of the peer review for fiscal year 1999, CRTD Vol. 56. New York: ASME; 1999.

ASME/RSI (The American Society of Mechanical Engineers/Institute for Regulatory Science). Assessment of technologies supported by the U.S. Department of Energy; results of the peer review for fiscal year 2000, CRTD Vol. 61. New York: ASME; 2000.

ASME/RSI (The American Society of Mechanical Engineers/Institute for Regulatory Science). Strategy for remediation of groundwater contamination at the Nevada Test Site; technical peer review report; report of the review panel, CRTD Vol. 62. New York: ASME; 2001a.

ASME/RSI (The American Society of Mechanical Engineers/Institute for Regulatory Science). Requirements for disposal of remote-handled transuranic wastes at the waste isolation pilot plant; technical peer review report; report of the review panel, CRTD Vol. 63. New York: ASME; 2001b.

ASME/RSI (The American Society of Mechanical Engineers/Institute for Regulatory Science). Assessment of technologies supported by the U.S. Department of Energy; results of the peer review for fiscal year 2001, CRTD Vol. 64. New York: ASME; 2001c.

ASME/RSI (The American Society of Mechanical Engineers/Institute for Regulatory Science). Waste isolation pilot plant initial report for polychlorinated biphenyl disposal authorization; technical peer review report; report of the review panel, CRTD Vol. 65. New York: ASME; 2002a.

ASME/RSI (The American Society of Mechanical Engineers/Institute for Regulatory Science). Airborne release fractions; technical peer review report; report of the review panel, CRTD Vol. 68. New York: ASME; 2002b.

ASME/RSI (The American Society of Mechanical Engineers/Institute for Regulatory Science). The beryllium oxide manufacturing process; technical peer review report; report of the review panel, CRTD Vol. 69. New York: ASME; 2002c.

ASME/RSI (The American Society of Mechanical Engineers/Institute for Regulatory Science). Assessment of technologies supported by the U.S. Department of Energy; results of the peer review for fiscal year 2002; CRTD Vol. 70. New York: ASME; 2002d.

ASME/RSI (The American Society of Mechanical Engineers/Institute for Regulatory Science). Radionuclide transport in the environment; technical peer review report; report of the review panel; CRTD Vol. 71. New York: ASME; 2002e.

ASME/RSI (The American Society of Mechanical Engineers/Institute for Regulatory Science). Review of selected nuclear safety programs at Savannah River Site; technical peer review report; report of the review panel, CRTD Vol. 72. New York: ASME; 2003a.

ASME/RSI (The American Society of Mechanical Engineers/Institute for Regulatory Science). Hanford Site 100 B/C Risk Assessment Pilot Project; technical peer review report; report of the review panel, CRTD Vol. 73. New York: ASME; 2003b.

ASME/RSI (The American Society of Mechanical Engineers/Institute for Regulatory Science). Salt waste processing facility technology readiness, CRTD Vol. 75. New York: ASME; 2003c.

ASME/RSI (The American Society of Mechanical Engineers/Institute for Regulatory Science). Spent Nuclear Fuel canister welding concept, CRTD Vol. 76. New York: ASME; 2004a.

ASME/RSI (The American Society of Mechanical Engineers/Institute for Regulatory Science). Spent Nuclear Fuel canister qualification support, CRTD Vol. 77. New York: ASME; 2004b.

Bechtel SAIC. Project functional and operational requirements, TDR-MGR-ME-000003, Rev. 2. Draft. Las Vegas, NV: Bechtel SAIC Company; (July) 2004.

Bechtel SAIC. Q-List. Rev. 0. TDRMGR-TR-000005. Prepared for U.S. Department of Energy, Office of Civilian Radioactive Waste Management, Office of Repository Development. Las Vegas, NV: Bechtel SAIC Company; (September) 2003.

Canning, J.J. Waste package closure system description document, 100-3YD-HW00-00100-000-001, Rev. 1. Las Vegas, NV: Bechtel SAIC Company; (March) 2004.

Canori, G.F.; Leitner, M.M. Project requirements document, TDR-MGR-MD-000001, Rev. 2. Las Vegas, NV: Bechtel SAIC Company; (December) 2003.

CRWMS M&O (Civilian Radioactive Waste Management System Management & Operating Contractor). Dose rate calculation for the 21 pressurized-water reactors uncanistered fuel waste package, CAL-UDC-NU-000002, Rev. 01. Las Vegas, NV: Civilian Radioactive Waste Management System Management & Operating Contractor; 2000.

Curry, P.M.; Loros, E.F. Performance objectives for the geologic repository operations area through permanent closure, PRD-002/T-012. Las Vegas; NV: Bechtel SAIC Company; (July) 2002a.

Curry, P.M.; Loros, E.F. Performance for the geologic repository after permanent closure, PRD-002/T-014. Las Vegas, NV: Bechtel SAIC Company; (July) 2002b.

DOE (U.S. Department of Energy). DOE handbook: Design considerations, DOE-HDBK-1132-99. Springfield, VA: U.S. Department of Commerce, Technology Administration, National Technical Information Service; (April) 1999. Available at http://tis.eh.doe.gov/techstds/standard/hdbk1132/hdbk113299.pdf

DOE (U.S. Department of Energy). DOE standard. Hoisting and rigging (formerly hoisting and rigging manual), DOE-STD-1090-2004. Washington, DC: U.S. Department of Energy; (June) 2004.

ESO (European Standard Organization). EN 292-1: Safety of machinery - basic concepts, general principles for design - part 1: Basic terminology, methodology. Bruxelles: European Committee for Standardization; (June 24) 1992a.

ESO (European Standard Organization). EN 292-2: Safety of machinery - basic concepts, general principles for design - part 2: Technical principles and specifications. Bruxelles: European Committee for Standardization; (June 24) 1992b.

ESO (European Standard Organization). EN 418: Safety of machinery - emergency stop equipment, functional aspects - principles for design. Bruxelles: European Committee for Standardization; (August 25) 1993a.

ESO (European Standard Organization). EN 775: Manipulating industrial robots - safety. Bruxelles: European Committee for Standardization; (August 25) 1993b.

ESO (European Standard Organization). EN 954-1: Safety of machinery - safety-related parts of control systems - part 1: General principles for design. Bruxelles: European Committee for Standardization; (May 8) 1997a.

ESO (European Standard Organization). EN1050: Safety of machinery - principles for risk assessment. Bruxelles: European Committee for Standardization; (October 23) 1997b.

Housely, G.K. Yucca Mountain weld cell design. Drawing 624935. Idaho Falls, ID: Idaho National Engineering and Environmental Laboratory; (September 4) 2003a.

Housely, G.K. Yucca Mountain weld closure facility drawings. Drawing 625198, 3-B. Idaho Falls, ID: Idaho National Engineering and Environmental Laboratory; (September 26) 2003b.

ICC (International Code Council). 2000 International building code. Falls Church, VA: International Code Council; 2000.

INEEL (Idaho National Engineering and Environmental Laboratory). Waste package closure system technical requirement, TFR-282. Idaho Falls, ID: Idaho National Engineering and Environmental Laboratory; (June 15) 2004a.

INEEL (Idaho National Engineering and Environmental Laboratory). Component design description: Welding and inspection system, TFR-283. Idaho Falls, ID: Idaho National Engineering and Environmental Laboratory; (July 29) 2004b.

INEEL (Idaho National Engineering and Environmental Laboratory). WPCS welding process: Control functions and associated performance requirements, EDF-5103. Idaho Falls, ID: Idaho National Engineering and Environmental Laboratory; (July 30) 2004c.

INEEL (Idaho National Engineering and Environmental Laboratory). Component design description: WPCS safety system, TFR-295. Idaho Falls, ID: Idaho National Engineering and Environmental Laboratory; (July 30) 2004d.

INEEL (Idaho National Engineering and Environmental Laboratory). Component design description: WPCS control and data management system, TFR-300. Idaho Falls, ID: Idaho National Engineering and Environmental Laboratory; (July 30) 2004e.

INEEL (Idaho National Engineering and Environmental Laboratory). Waste package closure system process flow diagram, EDF-4278. Idaho Falls, ID: Idaho National Engineering and Environmental Laboratory; (July 30) 2004f.

INEEL (Idaho National Engineering and Environmental Laboratory). Welding and inspection system configuration concept design decision for the Yucca Mountain repository waste package closure system, EDF-4227. Idaho Falls, ID: Idaho National Engineering and Environmental Laboratory; (July 21) 2004g.

INEEL (Idaho National Engineering and Environmental Laboratory). WPCS welding and inspection system radiation-hardened arc welding robot, SPC-572. Idaho Falls, ID: Idaho National Engineering and Environmental Laboratory; (July 30) 2004h.

INEEL (Idaho National Engineering and Environmental Laboratory). Interoffice work order, INEEL-TH-105043. Idaho Falls, ID: Idaho National Engineering and Environmental Laboratory; (March) 2003.

Kunerth, D.C. Ultrasonic Inspection design concept and development approach, EDF-4214. Idaho Falls, ID: Idaho National Engineering and Environmental Laboratory; 2004.

Lundin, C. Waste package welding processes selection consideration, MOL. 20021015.0209. Las Vegas, NV: Bechtel SAIC Company; (September 11) 2002.

McKerrow, P.J. Introduction to robotics. Boston, MA: Addison Wesley; 1991.

Minwalla, H.J. Project design criteria document, 000-3DRMGR0-00100-000-001, Rev 1. Las Vegas, NV: Bechtel SAIC Company; (March) 2003.

NFPA (National Fire Protection Association). NFPA 70. National electrical code. Quincy, MA: National Fire Protection Association; 2002a.

NFPA (National Fire Protection Association). NFPA 79. Electrical standard for industrial machinery. Quincy, MA: National Fire Protection Association; 2002b.

OSHA (Occupational Safety and Health administration). Occupational safety and health standards, 29 CFR 1910; 2003. Available at http://www.access.gpo.gov/nara/cfr/

RSI (Institute for Regulatory Science). Handbook of peer review. Columbia, MD: RSI; 2003.

Shelton-Davis, C.V. Quality program plan: Waste package closure system project, PLN-1626, Rev. 0, 05/06/04. Idaho Falls, ID: Idaho National Engineering and Environmental Laboratory; (May 6) 2004.

Shelton-Davis, C.V. Waste package closure system project execution plan, PLN-1678. Draft. Idaho Falls, ID: Idaho National Engineering and Environmental Laboratory; 2004.

USNRC (U.S. Nuclear Regulatory Commission). Disposal of high-level radioactive wastes in a geologic repository at Yucca Mountain, Nevada. Definitions, 10 CFR 63.2; 2003a. Available at http://www.access.gpo.gov/nara/cfr/

USNRC (U.S. Nuclear Regulatory Commission). Disposal of high-level radioactive wastes in a geologic repository at Yucca Mountain, Nevada. License application: Content of application, 10 CFR 63.21(c)(8); 2003b. Available at http://www.access.gpo.gov/nara/cfr/

USNRC (U.S. Nuclear Regulatory Commission). Disposal of high-level radioactive wastes in a geologic repository at Yucca Mountain, Nevada. Performance objectives for the geologic repository operations area through permanent closure, 10 CFR 63.111; 2003c. Available at http://www.access.gpo.gov/nara/cfr/

USNRC (U.S. Nuclear Regulatory Commission). Disposal of high-level radioactive wastes in a geologic repository at Yucca Mountain, Nevada. Performance objectives for the geologic repository after permanent closure, 10 CFR 63.113; 2003d. Available at http://www.access.gpo.gov/nara/cfr/

Van de Voorde, M.H.; Restat, C. Selection guide to organic materials for nuclear engineering, CERN 72-7. Genève: Centre Européen pour la Recherche Nucléaire (CERN); (May 17) 1972.

Vandergriff, K.U. Designing equipment for use in gamma radiation environments, ORNL/TM-11175. Oak Ridge, TN: Oak Ridge National Laboratory; (January) 1990.

Biographical
Summaries

Richard H. Adams is a Senior Engineer responsible for New Product Development at Central Research Laboratories in Red Wing, MN. He is responsible for designing and building prototype systems or components for customer evaluation, and for the design and integration of control systems; control system components; and programmable logic controller or imbedded software. He also develops and implements UL/CSA/CE testing requirements for mechanical and electrical systems. In addition, he develops functional requirements specifications and acceptance test specifications; and is involved with customer installation and acceptance testing. His recent activities include development of the Sterile Rapid Transfer Port system and the Stopper Lift system. Prior positions at Nuclear Technology in San Jose, CA, included Research Analyst, Engineering Manager, and Engineering Director. He managed the design and analysis of American Society of Mechanical Engineers Class MC pressure vessels and American Society of Mechanical Engineers Class I, II, and III piping systems. He was responsible for reviewing and certifying stress reports. As an Applications Analyst with Control Data Corporation in Arden Hills, MN, he was responsible for structural analysis, and for consulting with various clients using various Control Data engineering analysis products. While a Containment Engineer at Bechtel Power Corporation in San Francisco, CA, he was responsible for the development of a design specification for American Society of Mechanical Engineers Class MC pressure vessels and for administering their construction contracts. He is a member of the American Society of Mechanical Engineers and the International Society of Pharmaceutical Engineers. He has published various articles and is the recipient of two patents. Richard Adams received a B.S. degree in Mechanical Engineering and an M.S. degree in Engineering from the University of Washington in Seattle, WA. He is a registered Mechanical Engineer and a registered Civil Engineer in California.

Gary A. Benda is Vice President of Radiological Assistance, Consulting and Engineering, LLC (RACE) in Memphis, TN, specializing in radioactive waste processing and also President of U.S. Energy Corp. Previously, he was Vice President, General Manager of the Programs Division for NUKEM Nuclear Technologies, Inc. His responsibilities included developing and maintaining federal programs in North America that specialized in engineering and waste-processing services. Prior to NUKEM, he spent over 17 years with Chem-Nuclear Systems/WMX Technologies in various management roles. He also directed the site investigation, geophysical analysis, site screening, and license application; as well as managed the public hearings and licensing operations associated with local and national regulatory agencies for new low-level waste sites. He has over 20 years of experience in environmental restoration, technology development, and waste management; and has instructed over 20 national and international professional courses on radioactive waste management, mixed waste, and technology development. He is a member of the American Society of Mechanical Engineers (ASME), American Nuclear Society, and Health Physics Society. He has served as Chair of the ASME National Mixed Waste Committee; Environmental Remediation Committee; and Environmental Engineering Division. He has also chaired over 100 technical sessions at numerous national and international conferences on environmental management. He has authored and coauthored various scientific papers, reports, book chapters, and articles on the nuclear environment. Gary Benda is a Certified Health Physicist. He received a B.S. in Health Physics from Oklahoma State University, an M.S. degree in Applied Nuclear Science from Georgia Institute of Technology, and an M.B.A. from Seattle City University.

Erich W. Bretthauer is a consultant. Previously, he held the position of research professor at the University of Nevada-Las Vegas from January 1993 to March 1995. In that capacity, he served as Executive Director of Nevada Industry, Science Engineering & Technology, a public-private partnership which developed programs to enhance the scientific infrastructure of the state of Nevada. He was also the Assistant Administrator for Research and Development at the U.S. Environmental Protection Agency (EPA) from March 1990 until January 1993. In that capacity, he managed the Research and Development activities of a large and multi-disciplinary agency. Erich Bretthauer rose through the ranks of the EPA and served in a number of capacities ranging from a bench scientist to policy manager at national and international levels. He directed the EPA's emergency and long-term monitoring program after the accident at Three Mile Island, as well as its bioremediation program in Prince William Sound after the Valdez oil spill. Erich Bretthauer was the leader of the U.S. delegation and co-leader of a five-year North Atlantic Treaty Organization project which focused on exposures, risks, and measures to control dioxins. He also directed the EPA's ecological research program, and was Director of EPA's Environmental Monitoring Systems Laboratory in Las Vegas. He is a member of Sigma Xi; the American Chemical Society; and the American Water Works Association; and has served on the Federal Advisory Committee to the Civil Engineering Research Foundation. Erich Bretthauer is the author and coauthor of numerous papers, reports, and other publications. He received his B.S. and M.S. in chemistry from the University of Nevada, Reno, NV.

George E. Cook is currently Professor of Electrical Engineering, and Associate Dean for Research and Graduate Studies, Engineering School, Vanderbilt University, Nashville, TN. His studies focus on various aspects of robotic welding, including modeling and control of welding processes. He has also studied statistical process control applications to welding processes and Jacobian control for space manipulators. He started his career as an instructor; became an Assistant Professor; was promoted to Associate Professor; and eventually full professor, serving in various leading positions at Vanderbilt. He is a Fellow of the American Welding Society; a Fellow of the Institute of Electrical and Electronics Engineers; and has served as member of the Board of the American Welding Institute. He was also associated with a number of private corporations including CRC Welding Systems; Merrick Corporation; Advanced Control Engineering; and Industrial Electronics Laboratory. He has been awarded over 30 patents in the U.S. and foreign countries, and has over 200 publications. He has received numerous awards, including the James F. Lincoln Arc Welding Foundation Gold Award for the development of an adaptive robotic arc welding control methodology. George E. Cook received a B.E. degree from Vanderbilt University; an M.S. degree from the University of Tennessee; and a Ph.D. from Vanderbilt University, all in Electrical Engineering. He is a registered professional engineer in Tennessee, Kentucky, Alabama, and Wisconsin.

Ernest L. Daman is Chairman Emeritus of Foster Wheeler Development Corporation where he previously served as Director of Research and Chairman of the Board. He also held the position of Senior Vice President at the parent company, FWC. He is a Past President of American Society of Mechanical Engineers and was elected to the National Academy of Engineering. Ernest Daman is a Fellow of the Institute of Energy (England) and the American Association for the Advancement of Science, and Past Chairman of the American Association of Engineering Societies. He served on several American Society of Mechanical Engineers committees as member or chairman. Ernest Daman is the author of numerous papers and holds 18 patents. He was responsible for the design and development of a combined steam gas turbine plant, fluidized bed combustion, fast breeder reactor components, supercritical steam generators, environmental control processes, and advanced high-efficiency power generation systems. Ernest Daman received his B.M.E. degree from the Polytechnic Institute of Brooklyn.

Irwin Feller is Senior Visiting Scientist, American Association for the Advancement of Science and Professor Emeritus, Economics, Pennsylvania State University. His current research interests include the economics of academic research, and the evaluation of federal and state technology programs. He is the author of *Universities and State Governments*: A Study in Policy Analysis, and over 125 refereed journal articles; final research reports; book chapters; reviews; and numerous papers presented to academic, professional, and private organizations. He is former Chair of the Committee on Science, Engineering, and Public Policy, American Association for the Advancement of Science. Irwin Feller was the American Society of Mechanical Engineers Pennsylvania State Fellow from 1996 to 1997. He has been appointed to the National Research Council's Committee on Science, Engineering, and Public Policy; International Benchmarking of U.S. International Competitiveness-Immunology; Transportation Research Board, Research and Technology Coordinating Committee, National Research Council; and National Institute of Standards and Technology-Manufacturing Extension Partnership National Advisory Board. Irwin Feller is Chair, National Science Foundation's Advisory Committee on Social, Behavioral, and Economic Sciences and Chair, National Research Council's Committee on Assessing Behavioral and Social Science Research on Aging. He received a B.B.A. in Economics from the City University of New York and a Ph.D. in Economics from the University of Minnesota.

Robert A. Fjeld is Dempsey Professor of Environmental Engineering and Science at Clemson University. He coordinates the Department's nuclear environmental focus area, which is concerned with the environmental aspects of nuclear technologies including health physics, radioactive waste management, and risk assessment. Previously, he served as a faculty member in the Nuclear Engineering Department at Texas A&M University. He has active research on actinide transport in soils, instrumentation for measuring radioactivity in environmental samples, and environmental risk assessment. Robert Fjeld is a member of the Health Physics Society, American Nuclear Society, Society for Risk Analysis, and the American Society of Mechanical Engineers, where he serves as newsletter editor for the Mixed Waste Committee. He has served on two NRC Committees studying decontamination and decommissioning issues. Robert Fjeld is the author or coauthor of over 80 technical publications and presentations on topics such as radiation measurements, environmental transport of radionuclides, risk assessment, and aerosol physics. Robert Fjeld received a B.S. degree from North Carolina State University; and an M.S. degree and a Ph.D. from The Pennsylvania State University, all in Nuclear Engineering. He is a registered Professional Engineer.

William T. Gregory, III is currently Principal of Vinculum Marketing Solutions. Prior to forming Vinculum, he was Director of Government Programs for Foster Wheeler Environmental Corporation, an engineering and construction firm providing environmental and waste management services to government and private sector clients worldwide. Previously, he held a number of operational and business development positions at equipment manufacturing and service provision firms supporting nuclear utilities, industrial and process industries, and government agencies. His work has involved the management, processing, and disposition of hazardous, radioactive, and mixed wastes. He has also worked on the decontamination and decommissioning of nuclear facilities and on providing a wide range of environmental services in response to regulatory drivers. Prior to entering the private sector, he served with the U.S. Navy on nuclear submarines and at the operational command center for submarine operations in the Atlantic Fleet. William Gregory is actively involved with a number of international, national, and local organizations including the American Society of Mechanical Engineers and the American Nuclear Society. He is a founding member of the Board of Directors for the annual international Waste Management Symposium. William Gregory has served as an elected officer of several American Society of Mechanical Engineers divisions. He received a B.S. degree in Geology from the University of New Mexico, and an M.B.A. degree from Lamar University. He also attended naval nuclear power, nuclear weapons, and engineering schools as a U.S. Navy officer.

Nathan H. Hurt is a consultant in management and engineering with Technical and Management Consulting. He provides services to industrial firms and government agencies involved in environmental clean-up and waste management—both chemical and radioactive. He has extensive experience in the areas of executive management; plant management; engineering management; project management; marketing; and sales. He specializes in the areas of: uranium enrichment/production; engineering; development and marketing; plant management of rubber chemicals; petrochemicals; and thermoplastics. He also specializes in the engineering management of synthetic rubber and lattices; vinyl monomers and copolymers; polyesters; U.S. Department of Energy (DOE) weapons plants; quality assurance management; and operational readiness review. Nathan Hurt has been involved with the decommissioning of nuclear facilities. He was the Corporate Sponsor or Program Manager for seven decommissioning contracts at the DOE Complexes in Oak Ridge, TN; and Pinellas, FL. Previously, Nathan Hurt worked for Sharp and Associates, Inc. as the Director and Project Manager at the Oak Ridge Office. He was Vice President and Director of Oak Ridge Operations for IDM Environmental Corp., where he was responsible for the marketing and sales of decontamination, decommissioning, and waste management. He served as Project Manager for the laboratory quality assurance program at Westinghouse Hanford; DOE's Rocky Flats Plant—plant-wide identification of electrical equipment. He managed a study for a waste treatment and storage facility at the Portsmouth Area Uranium Enrichment Facility which included incineration and compaction of low-level radioactive wastes. He also worked for The Goodyear Tire and Rubber Company, including Goodyear Atomic, as Director of Research and Development, and President, where he was responsible for the operation of the Portsmouth Area Uranium Enrichment Facility. Nathan Hurt is a Past President of the American Society of Mechanical Engineers. He has been a member of: the American Association of Engineering Societies' Board of Governors; the American Institute of Chemical Engineers; and the Institute of Nuclear Materials Management. He is also a member of Tau Beta Pi Honorary Engineering Society; Pi Tau Sigma Honorary Mechanical Engineering Society; and was a member of The Nuclear Engineering Advisory Board of Worcester Polytechnic Institute. Nathan Hurt received a B.S. degree in Mechanical Engineering from the University of Colorado and has done Graduate, Technical, and Management course work at Pennsylvania State University. He is a registered Professional Engineer in Ohio.

Michael C. (M.C.) Kirkland is Vice President for the Southeastern Region of the Institute for Regulatory Science (RSI). In that capacity he leads various RSI projects related to the RSI mission in the southeastern U.S. Previously he was an independent consultant involved in peer review and various independent studies. For example, he led a team that performed an External Independent Review of the $1.3 billion Spallation Neutron Source Project at Oak Ridge, TN. He assisted in the planning and review of a management assessment at a U.S. Department of Energy (DOE) Site that involved the restart of a plutonium facility. He participated in planning, procurement, and review activities in the environmental remediation area that included decommissioning activities at a shut down nuclear test reactor; and designed and installed a ground water cleanup technology. M.C. Kirkland managed several environmental and construction projects that employed many soil investigative techniques including significant work with cone penetrometers. Additionally, he provided consulting services to a large environmental remediation services company regarding Dense Non-Aqueous Phase Liquid locating and removal techniques. During his tenure at the Savannah River Site (SRS) of DOE, M.C. Kirkland was a Technical Advisor, Project Manager, and Director of the Project Engineering Division. He evaluated nuclear and mixed waste conditions and aspects of high level wastes and spent nuclear fuel; determined material inventories; performed pollution prevention and environmental health and safety evaluations for a proposed waste treatment facility; served as technical advisor to a study administered by the Savannah River Operations Office; and developed integrated schedules defined for this project. M.C. Kirkland was director of the Project Engineering Division and managed the SRS design and construction program. He has been involved with waste management and environmental projects; cutting-edge technology programs; and worked with lasers and magnetic containment. He served as Director of the Waste and Fuel Cycle Technology Office, and planned and coordinated the programs of the DOE National High Level Waste Technology Office; the SR Fuel Cycle Technology Program; and the Commercial Interim Spent Fuel Management Program. M.C. Kirkland holds a B.S. degree in Mechanical Engineering from the University of South Carolina, and a Ph.D. in Management from Berne University in St. Kitts, West Indies. He is registered as a Professional Engineer in South Carolina.

Peter B. Lederman is a consultant with over 50 years of experience in all facets of process engineering, environmental management, operations, safety, control, and policy development. This includes hazardous substance management; environmental remediation; environmental audit; pollution prevention; development of air pollution control devices; and reuse of waste products. He recently retired as Executive Director of the Center for Environmental Engineering & Science; Executive Director for Patents and Licensing; and Research Professor of Chemical Engineering and Environmental Policy at the New Jersey Institute of Technology. Peter Lederman managed major programs in industrial waste treatment research and development, and in oil and hazardous material spill control and remediation. Most recently, he was responsible for a study of the Economic Impact of Environmental Regulations. He has been responsible for technology transfer efforts including the maturing and licensing of innovative environmental technologies. He is a Fellow and Director of the American Institute of Chemical Engineers; a Diplomat of the American Academy of Environmental Engineers; and a member of the ASME. He has served on several committees of the NRC and served as the chair of the NRC Committee on Review and Evaluation of the Army Chemical Stockpile Disposal Program. He is also a member of the NRC Committee on Chemical Weapons Disposal. He chaired the American Institute of Chemical Engineer's Environmental Division and chaired its Societal Impacts Operating Council. He is the recipient of a number of awards including the University of Michigan Alumni Medal of Merit Award in Chemical Engineering. Peter Lederman received a B.S.E., M.S.E., and Ph.D. (all in Chemical Engineering) from the University of Michigan in Ann Arbor, MI, and is a registered Professional Engineer.

Betty R. Love is currently Executive Vice President of the Institute for Regulatory Science. In that capacity, she is responsible for the management of day-to-day operations of the Institute, and for administration of several projects. She is the Administrative Manager of a large-scale peer review program in collaboration with the American Society of Mechanical Engineers for a number of organizations including the U.S. Department of Energy. Her current research activities center around the development and implementation of a systematic approach to stakeholder participation, notably in scientific meetings. Previously, Betty Love was Director, Department of Training and Information within the Office of Environmental Health and Safety of Temple University in Philadelphia, PA. During that period she was instrumental in the development of a "Handbook of Environmental Health and Safety". She also developed and implemented a large-scale training program not only for the faculty and staff of the University but also for others. Betty Love is currently Managing Editor of *Technology*. She has published several papers in peer-reviewed journals; has edited a number of compendia; and is the primary author of *Manual for Public and Stakeholder Participation*. Betty Love received a B.S. degree in Business Administration from Virginia State University in Petersburg, VA, and an M.S. degree in Developmental Clinical Psychology from Antioch College in Yellow Springs, OH.

Jeffrey A. Marqusee is currently the Technical Director of the Strategic Environmental Research and Development Program (SERDP), and the Director of the Environmental Security Technology Certification Program (ESTCP). SERDP is a tri-agency (U.S. Department of Defense [DOD], U.S. Department of Energy, and U.S. Environmental Protection Agency) environmental research and development program managed by the DOD. SERDP supports research and development to solve environmental issues of relevance to DOD in the areas of cleanup, compliance, conservation and pollution prevention. ESTCP is a DOD-wide program designed to demonstrate innovative environmental technologies at DOD facilities. ESTCP provides for rigorous validation of the cost and performance of new environmental technologies in cooperation with the regulatory and end-user communities. Prior to his current position, Jeffrey Marqusee served as a program manager for environmental technology in the Office of the Deputy Under Secretary of Defense for Environmental Security. He was the principal advisor to the Deputy Under Secretary on environmental technology issues. Before joining DOD, he worked at the Institute for Defense Analyses, where he advised both DOD and National Aeronautics and Space Administration in the areas of remote sensing, environmental matters and military surveillance. Jeffrey Marqusee has worked at Stanford University, the University of California and the National Institute of Standards and Technology. He has a Ph.D. in Physical Chemistry from the Massachusetts Institute of Technology.

A. Alan Moghissi is currently President of the Institute for Regulatory Science (RSI), a non-profit organization dedicated to the idea that societal decisions must be based on best available scientific information. The activities of the Institute include research, scientific assessment, and science education at all levels—particularly the education of minorities. Previously, Alan Moghissi was Associate Vice President for Environmental Health and Safety at Temple University in Philadelphia, PA and Assistant Vice President for Environmental Health and Safety at the University of Maryland at Baltimore. In both positions, he established an environmental health and safety program and resolved a number of relevant existing problems in those institutions. As a charter member of the U.S. Environmental Protection Agency (EPA), he served in a number of capacities, including Director of the Bioenvironmental/Radiological Research Division; Principal Science Advisor for Radiation and Hazardous Materials; and Manager of the Health and Environmental Risk Analysis Program. Alan Moghissi has been affiliated with a number of universities. He was a visiting professor at Georgia Tech and the University of Virginia, and was also affiliated with the University of Nevada and the Catholic University of America. Alan Moghissi's research has dealt with diverse subjects ranging from measurement of pollutants to biological effects of environmental agents. A major segment of his research has been on scientific information upon which laws, regulations, and judicial decisions are based—notably risk assessment. He has published nearly 400 papers, including several books. He is the Editor-in-Chief of *Technology: A Journal of Science Serving Legislative, Regulatory, and Judicial Systems*, which traces its roots to the *Journal of the Franklin Institute*—one of America's oldest continuously published journals of science and technology. Alan Moghissi is a member of the editorial board of several other scientific journals and is active in a number of civic, academic, and scientific organizations. He has served on a number of national and international committees and panels. He is a member of a number of professional societies including the American Society of Mechanical Engineers and is past chair of its Environmental Engineering Division. He is also an academic councilor of the Russian Academy of Engineering. Alan Moghissi received his education at the University of Zurich, Switzerland, and Technical University of Karlsruhe in Germany, where he received a doctorate degree in physical chemistry.

Lawrence C. Mohr, Jr., is currently Professor of Medicine, Biometry, and Epidemiology; and Director of the Environmental Biosciences Program at the Medical University of South Carolina. His areas of research and special interest include internal medicine, pulmonary disease, environmental medicine, risk assessment, molecular epidemiology, and biomarker applications. Prior to assuming his current position, Mohr served on the medical staffs at Walter Reed Army Medical Center and the George Washington University Medical Center in Washington, D.C. During this time, he also served as a physician in the White House Medical Unit—which provides medical support to the President of the United States. He has held previous academic appointments at the Uniformed Services University of the Health Sciences in Bethesda, Maryland, and at George Washington University in Washington, DC. In addition to directing a nationally-prominent research program in the environmental health sciences, he lectures throughout the world and has held visiting professorships at multiple universities. Lawrence Mohr has served on numerous scientific, professional, and government boards and committees. He is a member of several professional societies including the American Federation for Medical Research; the Society for Risk Analysis; the Society of Medical Consultants to the Armed Forces; and the Wilderness Medical Society. He also is a Diplomat of the

American Board of Internal Medicine and is a Fellow of both the American College of Physicians and the American College of Chest Physicians. He has authored or coauthored more than 80 articles, books, or technical publications. Lawrence Mohr received an A.B. degree in Chemistry with highest honors as well as an M.D. degree, both from the University of North Carolina, Chapel Hill. His postdoctoral education includes an internship and residency in Internal Medicine and fellowship training in Pulmonary Medicine, all at Walter Reed Army Medical Center in Washington, D.C.

Goetz K. Oertel's career in engineering, physics, chemistry, astronomy, and technical program management spans more than 40 years. He consults for industrial, academic, and governmental organizations in North and South America. As President and CEO of the Association of Universities for Research in Astronomy, a nonprofit corporation, he engineered the initiation and completion of two 8-m aperture optical telescopes, and oversaw the Space Telescope Science Institute from before launch, through repair of the "Hubble flaw", to its successful operation. He initiated the first study of the Next Generation Space Telescope that will succeed Hubble, proposed the Advanced Solar Telescope, and he oversaw the completion of ambitious ground-based astronomy facilities. He held technical and management positions in the U.S. Department of Energy, including Director of Defense Waste Management; Acting Manager of the Savannah River Operations Office; Deputy Manager of Albuquerque Operations Office; and Deputy Assistant Secretary for Safety, Health, and Quality Assurance. He had primary responsibility for the congressionally-mandated Defense Waste Management Plan, and for managing the related technology development, operations, and projects. He led the initiation of the Defense Waste Processing Facility, and saw it and the Waste Isolation Pilot Plant through technical, managerial, stakeholder, and political challenges. He was National Aeronautics and Space Administration Space Science Chief and Program Manager, and Aerospace Engineer at Langley. He was selected to a government-wide executive development program and served in the White House with the President's Science Advisor and in the Office of Management and Budget's Space and Energy branch. He chaired the Westinghouse West Valley Corporation Technical Advisory Group for high-level nuclear waste vitrification and management before, during, and after that project's successful vitrification campaign. He was appointed as Associate Member for life of the National Academies. He is a member of the American Physical Society, Sigma Xi, and other professional organizations. He was elected Fellow of the American Association for the Advancement of Science. He is Chair or member of boards and committees of the National Research Council; George Mason University; the Center of Excellence in Hazardous Materials Management; the American Society of Mechanical Engineers; International University Exchange; and Westinghouse West Valley Corporation. He is a founding member of the Editorial Board for "Data Science", the new international on-line journal of Codata. He published numerous peer-reviewed papers and was awarded two patents. Trained as physicist and chemist, he received a Vordiplom in Physics and Chemistry from the University of Kiel while on German industrial and governmental scholarships, and a Ph.D. in Physics from University of Maryland at College Park under a Fulbright scholarship.

Francis J. Patti is currently an independent consultant with expertise in the fields of Civil Engineering and Nuclear Engineering. He was Chief Nuclear Engineer at Burns & Roe (B&R). He has been involved in nine major nuclear power projects, and has also worked on hot cell facilities; research reactors; power reactors; radioactive waste facilities; and decommissioning projects. His work covered the full range of architect-engineer activities and includes both analytical work and systems designs. His consulting assignments at B&R have included working with Princeton Plasma Physics Laboratory on the Tokomak Fusion Test Reactor; Duquesne Light Company on Beaver Valley 2; Public Service of New Hampshire on Seabrook; Gulf States Utilities on the River Bend Nuclear Plant; Toledo Edison on the Davis-Besse Nuclear Power Station; Korea Power Engineering Company; North Carolina Electric Membership Corporation on Catawba Nuclear Units 1 and 2; and the Brookhaven National Laboratory (BNL). At BNL, he was project leader on engineering a major modification to the Brookhaven Medical Research Reactor—to enhance its capability for boron neutron capture therapy to treat certain types of brain cancer. His research reactor experience includes major participation in the design of the Argonne Advanced Research Reactor; the Army Pulsed Reactor Facility; and the NBS Research Reactor. While on these assignments, Francis Patti reviewed, evaluated, or was involved with: providing welding engineering support at construction sites; air modeling, permitting, and hazardous waste issues; waste water treatment systems for power plants; emergency plans; procedures; safety analysis; tritium systems upgrade engineering; architect-engineering capabilities; and the status of piping verification work. Since his retirement from B&R, Francis Patti's consulting assignments have also involved preparing material for their defense in an asbestos lawsuit, the emergency spray pond for the American nuclear plant and investigating conversion of the Rostov nuclear plant to fossil fuel firing. The latter involved substantial contact with United Engineers & Constructors. Francis Patti has served as Chair of the local sections of the American Nuclear Society and the American Society of Civil Engineers. He received Civil Engineer of the Year and Distinguished Service Awards from the local branch of the American Society of Civil Engineers. He has also served on several peer review panels for the American Society of Mechanical Engineers, which dealt mostly with treatment and disposal of radioactive waste. He is the author or coauthor of 15 technical papers. He holds a B.S. degree in Civil Engineering from Drexel University, Philadelphia, PA; and an M.S. degree in Civil Engineering from the Massachusetts Institute of Technology, Cambridge, MA. He also has a diploma in Nuclear Science and Engineering from the International School of Nuclear Science and Engineering, as well as M.S. and Professional Degrees in Nuclear Engineering from Columbia University, New York, NY. Francis Patti is a registered Professional Engineer in New York.

Sorin R. Straja is currently Vice President for Science and Technology of the Institute for Regulatory Science. He has over 20 years of expertise in mathematical modeling and software development as applied in chemical engineering and risk assessment. Previously he served as Assistant Professor of Biostatistics with Temple University, Philadelphia; as Director of the Department of Occupational Health and Safety of Temple University, Philadelphia; and as a chemist with University of Maryland at Baltimore. Sorin Straja has extensive experience in the chemical industry where he worked as a senior R&D consultant with the Chemical and Biochemical Energetics Institute, and as a plant manager with Chemicals Enterprise Dudesti and Plastics Processing Bucharest from Romania. He was an Assistant/Adjunct Professor of Chemical Engineering with the Polytechnic Institute Bucharest. Sorin Straja is the author of two books and 44 scientific papers published in internationally recognized and peer-reviewed journals. He was an editor of *Environment International*, and currently is a contributing editor of *Technology*. Sorin Straja received a Certificate of Appreciation for Teaching from Temple University, the "Nicolae Teclu" Prize of the Romanian Academy, and a Certificate of Appreciation from U.S. Department of Agriculture for significant volunteer contributions. He is a Fellow of the Global Association of Risk Professionals, and a member of the American Chemical Society, American Institute of Chemical Engineers, Society for Risk Analysis, and New York Academy of Sciences. Sorin Straja holds a M.S. in Industrial Chemistry and a Ph.D. in Chemical Engineering both from Polytechnic Institute Bucharest.

Glenn W. Suter, II is currently Science Advisor at the National Center for Environmental Assessment of the U.S. Environmental Protection Agency (EPA) in Cincinnati, OH. Previous to his current position, he was at Oak Ridge National Laboratory, initially as Research Associate and gradually rising to Science Leader at the Environment Science Division of the Laboratory. His interest has focused on Ecotoxicology in general and Ecological Risk Assessment in particular. He is one of the developers of the most widely-used methodology for Ecological Risk Assessment. This method has been applied to the impact of pollutants on fish, contaminated soils, production of synthetic fuels, and various other ecosystems. Glenn Suter has lectured widely, both nationally and internationally on Ecological Risk Assessment. He is currently a member of the U.S. EPA's Risk Assessment Forum. He has been a member of numerous panels and has consulted with various governmental agencies and private organizations, including the Council of Environmental Quality. He was a member of the Scientific Review Panel for Savannah River Ecology Laboratory; the National Science Foundation Panel on Decision Making and Valuation for Environmental Policy; and the U.S. EPA Science Advisory Board and Conservation Foundation, Ecosystem Valuation Forum. In addition, he was a member of the International Institute of Applied Systems Analysis Task Force on Risk and Policy Analysis and the Council on Environmental Quality. He was a member of the Board of Directors, for the Society for Environmental Toxicology and Chemistry. Glenn Suter is presently on the Editorial Board of *Environmental Health Perspectives* and *Human and Ecological Risk Assessment*. Previously, he was on the Editorial Board of *Handbook of Environmental Risk Assessment and Management* and *Environmental Toxicology and Chemistry*. Glenn Suter is the author of three books and is author and coauthor of over 200 publications. He received a B.S. degree in Biology from Virginia Polytechnic Institute and a Ph.D. in Ecology from the University of California, Davis.

Karyanil T. Thomas has more than 45 years of professional experience in the nuclear field with extensive accomplishments in nuclear waste management and disposal, and environmental management technologies. His areas of specialization are chemical/nuclear engineering and sciences, and radioactive waste management. After retiring from the National Research Council in 1998, he worked as a consultant to the International Atomic Energy Agency and the Board on Radioactive Waste Management of the National Research Council on specific projects. Karyanil Thomas directed two major projects for the U.S. Department of Energy (DOE) at the National Research Council-Board on Radioactive Waste Management, one of which dealt with the feasibility and potential applications of transmutation technologies using advanced reactors and accelerators for the safe disposal of high-level radioactive waste. The report that evolved from this study is considered to be one of the most authoritative documents on this highly-complex subject. The second study focused on environmental management technologies which reviewed the DOE's Office of Science and Technology's technology development programs for the cleanup of nuclear weapons complex facilities. Prior to joining the National Research Council, he was responsible for the radioactive waste disposal program of the International Atomic Energy Agency covering technical, regulatory, and safety aspects for all categories of wastes produced in the nuclear industry. Karyanil Thomas was involved with the development of the nuclear program in India—which consisted of the production of uranium and thorium to radioactive waste management in the nuclear fuel cycle—from its infancy. He implemented radioactive waste management systems for research and nuclear power reactors; fuel fabrication and reprocessing facilities; and related research and development activities. He was also responsible for directing the site selection and the design and construction of repositories for low- and intermediate-level radioactive wastes produced in the Indian nuclear industry. In addition, he headed a study which prepared a major report on the feasibility of setting up large, integrated, dual-purpose nuclear power reactors with desalination and fertilizer plants for the production of power; water; and phosphatic and nitrogenous fertilizers through energy-intensive processes. He is a past associate member of the American Institute of Chemical Engineers; and past executive member and Fellow of the Maharashtra Academy of Sciences (India). He was also a member of the Advisory Board for the Central Salt and Marine Research Institute in India. He is author and coauthor of more than 68 publications in the nuclear and chemical sciences and engineering fields with particular reference to radioactive waste management and disposal. Karyanil Thomas received a B.S. degree in Technology from the Benares Hindu University, Benares, India; and an M.S. degree in Chemical Engineering from North Carolina University.

Cheryl A. Trottier is currently Chief of the Radiation Protection, Environmental Risk, and Waste Management Branch of the Office of Nuclear Regulatory Research at the U.S. Nuclear Regulatory Commission.

In that capacity, she is responsible for the management of research programs and the development of technical bases to support rule-making. This includes the development of models for realistically assessing the radiation doses to the public that are likely to be received from lands and materials cleared from regulatory control; evaluating hydrologic model and parameter uncertainty; the development of realistic parameters for assessing sorption processes in geochemical models; and refining evaluations of radionuclide transport mechanism in the environment. In her 30 years of experience in the field of radiation protection, she has been involved in the management of environmental radiation protection monitoring programs and laboratory measurements, and the emergency preparedness coordination for an electric utility. She was also involved in the areas of materials use regulation oversight; development of regulations; and the development of guidance for use of byproduct and special nuclear materials. This included finalization of regulations and the development of regulations to certify the gaseous diffusion plants. Cheryl Trottier serves as one of the U.S. representatives to the Nuclear Energy Agency Committee on Radiation Protection and Public Health. She is a member of the American Nuclear Society, and serves as a member of the Committee on Site Clean-Up Restoration Standards. She received her B.A. degree in Biology from Rutgers University.

Charles O. Velzy is a consultant in the field of waste treatment and disposal. Previously, he held increasingly responsible positions with the environmental consulting engineering firm, Charles R. Velzy Associates, Inc., becoming President in 1976. In 1987, when Velzy Associates merged with Roy F. Weston, Inc., Charles Velzy became Vice President of Weston, a position which he held until retiring in 1992. He has over 35 years of experience as an environmental engineering consultant specializing in: the analysis of waste management problems; design of wastewater treatment and waste disposal systems; and design of new, retrofit of existing, testing, and permitting of waste combustion facilities. He has authored or co-authored over 80 publications—primarily in the field of solid waste management. He has served on the Science Advisory Board of the U.S. Environmental Protection Agency; as President of the American Society of Mechanical Engineers (ASME); Chair of the ASME Peer Review Committee; and as Treasurer of the American Academy of Environmental Engineers (AAEE). He has served on numerous committees of the ASME, the AAEE, the American National Standards Institute, and the American Society for Testing and Materials. He is a registered professional engineer in New York and eleven other states. Charles Velzy received B.S. degrees in Mechanical and Civil Engineering, and an M.S. in Sanitary Engineering from the University of Illinois at Urbana-Champaign.

Acronyms

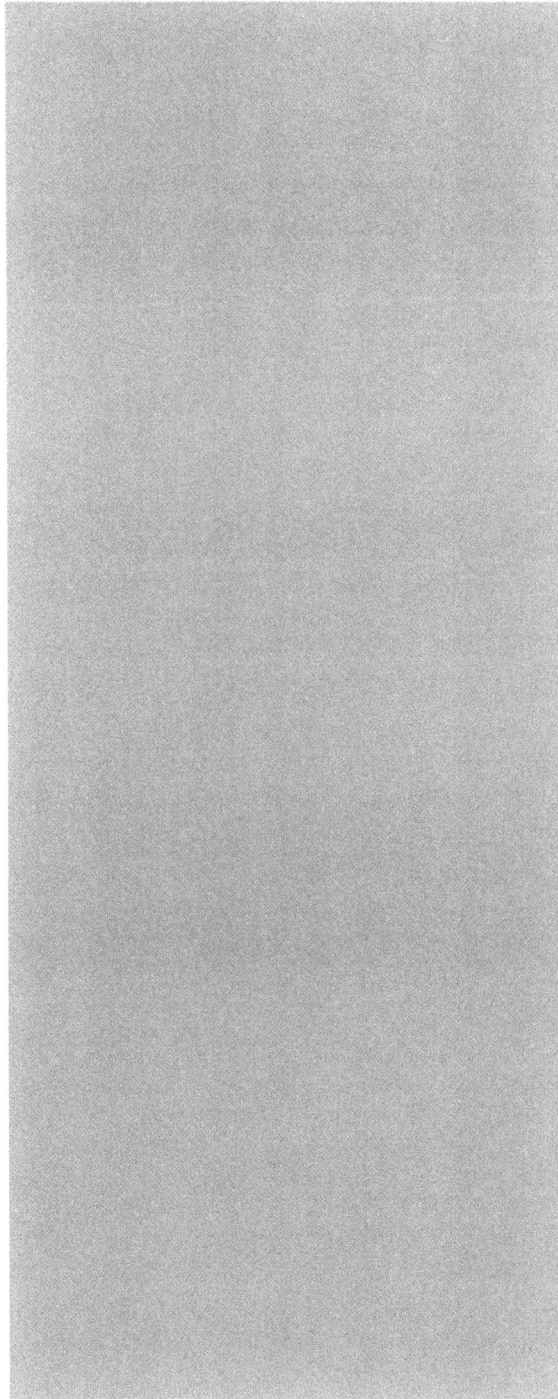

ALARA	as low as reasonably achievable
ASME	American Society of Mechanical Engineers
ASTM	American Society of Testing and Materials
AVC	automatic voltage control
BSC	Bechtel SAIC Company, LLC
C&DMS	WPCS Control and Data Management System
CDD	Component Design Description
CE	Consumer Electronics Association
CFR	Code of Federal Regulations
COTS	commercial off the shelf
DCMIS	Digital Control and Management Information System
DD&D	deactivation, decontamination, and decommissioning
DOE	U.S. Department of Energy
EP	Executive Panel
ESO	European Standard Organization
ET	eddy current testing
F&OR	Functional and Operational Requirement
GUI	graphical user interface
HDCM	hardware device control module
HDM	hardware device module
HHIT	HDCM/HDM Interface Tool
HLW	high-level waste
HMI	human-machine interface
HVAC	heating, ventilating, and air conditioning
ICS	Industrial Control and Systems
INEEL	Idaho National Engineering and Environmental Laboratory
LSR	licensed specification requirement
MGR	Monitored Geologic Repository
MSM	master-slave manipulators
MTBF	Mean time between failures
MTTR	mean time to repair
NDE	non-destructive examination
NEMA	National Electrical Manufacturers Association
NFD	no further decomposition (of requirement within the document)
NFPA	National Fire Protection Agency
NTSC	National Television System Committee
O&M	operation and maintenance
OW	operator workstation
PC	personal computer
PEEK	polyetheretherketone
PID	proportional integrative derivative
PLC	programmable logic controller
PRCEE	Peer Review Committee for Energy and Environment
PRD	Project Requirements Document
PT	Project Team
RHS	remote handling system
RP	Review Panel

RSI	Institute for Regulatory Science
SCADA	Supervisory Control and Data Acquisition
SCS	Supervisory Control System
SNF/HLW	spent nuclear fuel/high-level waste
SO	system operability (acceptance test)
SSC	structures, systems, and components
TBD	to be determined
TCP/IP	Transmission Control Protocol/Internet Protocol
TFR	Technical and Functional Requirements
UL	Underwriters Laboratories, Inc.
USEPA	United States Environmental Protection Agency
USNRC	United States Nuclear Regulatory Commission
UT	ultrasonic testing
VT	visual testing
WP	waste package
WPCA	waste package closure area
WPCS	Waste Package Closure System
WPID	waste package identification
YMP	Yucca Mountain Project

Definitions

Closure cell (operating floor)	All closure operations are performed in an operating-floor closure cell, which lines up with the waste package secured on the ground floor.
Closure maintenance area (operating floor)	Crane maintenance is performed, and some materials are stored, in this area. A radiation wall separates the area from the closure cell, which has a sliding portion for access into the closure cell. Cranes can pass above the closure maintenance area, radiation wall, and closure cell. The waste package closure maintenance area is equipped with air locks to pass equipment and provide personnel access when required from the closure support area.
Closure operating gallery (operating floor)	The closure operating gallery is located on the opposite side of the closure support area, with the closure cells in the middle of the two areas. The process operators are located in this area. Windows are located in the wall separating the closure operating gallery and the closure cell to view the operations. It is designed to allow for continuous occupancy for personnel supporting the closure cell operations.
Closure support area (operating floor)	This area supports the closure cell with glovebox operation and the transfer of lids, consumables, and tools required for the closure process. It is located adjacent to the closure cell with a shielding/contamination wall between the two areas. It is designed to allow for continuous occupancy for personnel supporting the closure cell operations.
Component	Item of equipment such as a pump, valve, or relay; or an element of a larger array such as computer software, length of pipe, elbow, or reducer.
Deactivation, decontamination, and decommissioning	Generally refers to the set of activities or phase of the project dealing with the final disposition of the facility, for example, permanently disabling or de-energizing equipment, final decontamination (if necessary), and dismantlement for reuse or disposal.
Digital Control and Management Information System	The facility software control system that directs the movement of a waste package from the time it enters a facility until it is loaded out of the facility. It interfaces with the WPCS to provide information essential to the waste package closure process.
Facility	The Facility is a general term referring to the Yucca Mountain Project surface buildings where spent nuclear fuel and high-level waste is received, loaded into a waste package, sealed, and prepared for emplacement in the repository.
Functional requirement	This requirement specifies what the system must do.
HHIT	HDCM/HDM Interface Tool. This device interfaces power, signal, control and utilities between an hardware device control module and a hardware device module. It performs as an umbilical to the module in the closure cell.
Layup	A period, not a process, during which the Facility is monitored and maintained in a stable and known condition. Note that this term is synonymous with the terms *surveillance* and *maintenance* in standard.

Maintenance area (upper floor)	The maintenance area is located directly above the closure cell, closure support area, and waste package closure maintenance areas. It is used for placement of major equipment that does not need to be in the closure cell (such as power supplies, vacuum pumps, and spare equipment). Access is gained by a freight elevator or through the upper floor hatches into the closure cells and the closure support area. It is designed for continuous occupancy for personnel supporting closure cell operations.
Operational requirement	A requirement that specifies how well the system must operate.
SCADA	SCADA is not a full control system, but rather focuses on the supervisory level. As such, it is a purely software package that is positioned on top of hardware to which it is interfaced, in general via Programmable Logic Controllers (PLCs), or other commercial hardware modules.
Shutdown (also safe shutdown)	The set of activities (i.e., process) performed to mitigate facility hazards and to place said Facility in stable and known conditions that are cost-effective to maintain. Shutdown may also be used to describe the state of the Facility after shutdown activities were successfully performed. Note: This term is related to the term deactivation in the standard deactivation, decontamination, and decommissioning (DD&D) vernacular, which implies permanent disabling of equipment. However, as used in this plan, shutdown relative to equipment and systems implies temporary versus permanent disabling or de-energizing (e.g., disconnecting equipment from its source of power by an easily reversible method). Deactivation as a part of DD&D has a more permanent connotation.
Structure	Elements that provide support or enclosure, such as buildings, freestanding tanks, basins, dikes, and stacks.
Subsystem	The Waste Package Closure System comprises about a dozen subsystems, such as welding and inspection, leak detection, material tracking, etc.
System	The word *system* refers to the Waste Package Closure System (WPCS) unless otherwise specifically denoted, such as in "supervisory control system."
Waste package closure area (WPCA)	The area within the Facility where closure operations will be performed, consisting of three floors. The ground floor is where the waste package enters and is located in the waste package stations. The operating floor contains the closure cell, closure support area, closure operating gallery, and closure maintenance areas. The upper floor contains the maintenance area.
Waste package closure system (WPCS)	The integrated set of subsystems, components, and equipment that form the operational capability to close and seal the waste package.
Waste package positioning cell (ground floor)	This positioning cell is located on the ground floor under the waste package closure area. The waste package enters from the fuel transfer area and is docked in one of the waste package position cells. This area is not part of the WPCS scope.

www.ingramcontent.com/pod-product-compliance
Lightning Source LLC
Chambersburg PA
CBHW082035230326

41598CB00081B/6516